Acclaim for S ... 's

HOW WE DIE

"The story comes from a sensitive observer . . . who has seen much, taken much thought, and written it all down with a superior gift for descriptive narrative. . . . Nuland's meditations . . . reflect a pragmatic, measured skepticism permeated by an intensely human compassion." —*Washington Post Book World*

"Powerfully eloquent . . . a relentlessly frank and graphic description of the ways human life achieves its terminus."
—*The New York Times*

"As powerful and sensitive, and unsparing and unsentimental as anything I have ever read." —Oliver Sacks

"Nuland proposes what almost anyone who has been touched by death will recognize as common sense. . . . [He] never shies away from the cultural implications of his profession. . . . You cannot read *How We Die* without becoming aware of your body, if only . . . to ask it impermissible questions. You put the book down merely to pick it up again." —*The New Yorker*

"Stunningly frank . . . strips away the fantasy about what happens when the body stops." —*Newsweek*

"Any reader who is not still convinced of his own mortality . . . is bound to be altered in some profound way by this book. . . . This is knowledge we all should have." —*USA Today*

"Nuland combines the clinical eye of a trained physician with . . . emotional and philosophical reflectiveness." —*Newsday*

Sherwin B. Nuland

HOW WE DIE

Sherwin B. Nuland, M.D., is the author of *Doctors: The Biography of Medicine.* He teaches surgery and the history of medicine at Yale, and is also Literary Editor of *Connecticut Medicine* and Chairman of the Board of Managers of the *Journal of the History of Medicine and Allied Sciences.* In addition to his numerous articles for medical publications, he has written for *The New Yorker, The New Republic,* and *Discover* magazine. His book *The Origins of Anesthesia* is a volume in the Classics of Medicine Library. Dr. Nuland and his family live in Hamden, Connecticut.

Also by Sherwin B. Nuland

The Origins of Anesthesia

Doctors: The Biography of Medicine

Medicines: The Art of Healing

The Face of Mercy

HOW WE DIE

HOW WE DIE

Reflections on Life's Final Chapter

Sherwin B. Nuland

VINTAGE BOOKS
A Division of Random House, Inc.
New York

FIRST VINTAGE BOOKS EDITION, JANUARY 1995

 Published in the United States by Vintage Books, a division of Random House, Inc., New York, and simultaneously in Canada by Random House of Canada Limited, Toronto.Originally published in hardcover by Alfred A. Knopf, Inc., New York, in 1994.

The Library of Congress has cataloged the Knopf edition as follows:

Nuland, Sherwin B.
How we die / Sherwin B. Nuland.
p. cm.
Includes index.
ISBN: 0-679-41461-4
1. Death. I. Title.
BD444.N85 1994
616.07'8—dc20 93-24590
CIP

Vintage ISBN: 0-679-74244-1

Manufactured in the United States of America

3 5 7 9 C 8 6 4

To my brothers,
Harvey Nuland and Vittorio Ferrero

. . . death hath ten thousand several doors
For men to take their exits.

—John Webster, *The Duchess of Malfi*, 1612

Contents

Acknowledgments

The eighteenth-century novelist Laurence Sterne once remarked that writing "is but a different name for conversation." The content and tone of a book or essay are determined by the author's perception of the reader's anticipated response to each sentence as it is given form on the page—the reader is always present. The book you are about to read was conceived with no other plan in mind than that of conversing with people who want to know what it is like to die. I have tried to hear how a reader might reply to what is being said. By listening well, I hoped to be able to address every response as immediately and clearly as possible.

The dialogue in these chapters, however, is only the culmination of other conversations I have been having most of my life—with my family, my friends, my colleagues, and above all my patients—with those who have been closest to me and whose wisdom I have sought in order to come to an understanding of what our lives, and our deaths, are about. To seek wisdom in another's words is much less difficult than to find it in another's experience. I have looked for it everywhere I thought it could be discovered. Even when I had no idea I was learning from one or another of the vast number of men and women whose lives have entered mine, they were nevertheless teaching me, usually with equal unawareness of the gift they were bestowing.

Although most learning is thus subtle and unrecognized as such by either its recipients or its providers, a great deal of it does grow out of the more usual kind of conversation: direct verbal interchange between two people. In my own case, the most extensive of those dialogues have gone on intermittently for years or even decades, while a few have taken place only during the writing of the book. If "conference [maketh] the ready man" as Francis Bacon claimed, then I have been made ready for *How We Die* by countless hours in the company of extraordinary people.

Several of my fellow members of the Bioethics Committee at the Yale–New Haven Hospital have again and again sharpened my comprehension of critical issues faced not only by patients and health professionals but at one time or another by all of us. I am particularly indebted to Constance Donovan, Thomas Duffy, Margaret Farley, Robert Levine, Virginia Roddy, and Howard Zonanna. Together and individually, they have shown me an image of medical ethics that is as humane (and even spiritual) as it is intellectually disciplined.

Thanks go also to another member of the committee, Alan Mermann, a pediatrician who found renewed vigor as a Congregationalist minister and the chaplain of our medical school. He has been generous in helping me understand what it is like for medical students and dying patients to befriend each other and share one another's fears and hopes.

Ferenc Gyorgyey has made available the vast resources of the historical collections at Yale's Cushing/Whitney Library, but his even greater gift during these many years has been the equally vast resources of his friendship and his wide-ranging intellect. Jay Katz, both in our conversations and in his writings, has taught me a sensitivity to medical decision-making that transcends the mere clinical facts of a patient's illness and even the conscious motivations that would seem to determine choice of treatment options. My wife, Sarah Peterson, teaches me yet another kind of sensitivity, which is sometimes called charity and sometimes called love. In charity or love there is an understanding of another's perceptions and there is also unquenchable faith. In Sarah's tradition: "Though I speak with the tongues of men and of angels and have not love, I become as a sounding brass or a tinkling cymbal." Therein lies a great lesson not only for individuals but for nations and professions—especially my own profession of medicine.

For the past decade, I have benefited from the friendship of Robert Massey. As a practicing internist, a medical school dean, and a historian of medicine as well as a commentator on its present and future, Bob Massey has transmitted to several generations of his physician colleagues a dimension of understanding and a sense of medical obligation that surpass the ephemeral concerns of the moment and the parochial concerns of the guild. I have taken advantage of his friendship by making him my sounding board, my oracle, and even my authority for classical allusions, not to mention Latin grammar. There is almost nothing in this book that he and I have not discussed. His confidence in the value of this undertaking has been a source of quiet energy for me over these many months of work.

Each chapter of *How We Die* has been reviewed by one or more authorities on its content. In every case, important suggestions have resulted from the readings which contributed in significant ways to my ability to clarify the material. The cardiac chapters were critiqued by Mark Applefeld, Deborah Barbour, and Steven Wolfson; the sections on aging and Alzheimer's disease by Leo Cooney; the trauma and suicide section by Daniel Lowe; the AIDS chapters by Gerald Friedland and Peter Selwyn; the clinical and biological aspects of cancer by Alan Sartorelli and Edwin Cadman; the discussion of the doctor-patient relationship by Jay Katz. Specialists in these areas will easily recognize the names of each of my consultants—I honor myself by recording them here. They have been generous beyond my expectation.

A number of people have helped me answer specific queries and track down sources: Wayne Carver, Benjamin Farkas, Janis Glover, James M. L. N. Horgan, Ali Khodadoust, Laurie Patton, Johannes van Straalen, Mary Weigand, Morris Wessel, Ann Williams, Yan Zhangshou, and my great-hearted secretary, Rafaella Grimaldi. G. J. Walker Smith reviewed an autopsy series with me and helped to put its findings into the context of the degenerative processes of aging. A morning spent with Alvin Novick opened my eyes to political and intensely personal aspects of AIDS that I had only guessed at—it could not have been easy for Al to expose to a virtual stranger the pain in his still-grieving heart, but somehow he found the strength to do it, and I will not forget what he taught me. Irma Pollock, whom I have admired since childhood, spoke to me through the anguish of recalling the tragedy of Alzheimer's disease, because she wanted to help others. Her story has strengthened my faith in the power of selfless love.

The entire text of *How We Die* was read by several people of disparate backgrounds, whose comments proved extremely helpful in my own final scrutiny: Joan Behar, Robert Burt, Judith Cuthbertson, Margaret DeVane, and James Ponet. It goes without saying that Bob Massey and Sarah Peterson made numerous critical contributions as they reviewed the evolving work, chapter by chapter. Bob's style is benevolent and diplomatic, but that Peterson woman is unsparing in her pursuit of what I have elsewhere called "the recognition of rambling and the discouragement of drift." I always made the changes when she pointed them out—even *her* charity has its limits.

And finally, to my new friends in the world of publishing. *How We Die* originated in the vision of Glen Hartley—not only the idea but even the title was his. At Dan Frank's suggestion, he and Lynn Chu sought me out and presented me with a mission I could not turn away from. The manuscript that ultimately resulted was passed through the filter of Dan's skillful editorial mind; only his authors can fully appreciate the value of such guidance. Sonny Mehta carried this project in his own gentle hands from inception to conclusion, as its editor, publisher, and chief booster. If there is an all-star team in publishing, this must surely be it.

It is said that in the twentieth century there are no longer any Muses, but I have found one. Her name is Elisabeth Sifton, and I have tried to treat ideas and the English language in ways that will please her. I ask no greater reward than her approval.

There is a second of Laurence Sterne's aphorisms that applies to *How We Die*. It is this: "Every man's wit must come from every man's soul, and no other body's." This is my book. No matter the inspiration and contributions of so many others, I declare every bit of it—every conception and every misconception, every truth and every error, every helpful thought and every useless interpretation—to be my own. They are no other body's. *How We Die* is no other body's because this book comes from my soul.

S.B.N.

Introduction

Everyone wants to know the details of dying, though few are willing to say so. Whether to anticipate the events of our own final illness or better to comprehend what is happening to a mortally stricken loved one—or more likely out of that id-borne fascination with death we all share—we are lured by thoughts of life's ending. To most people, death remains a hidden secret, as eroticized as it is feared. We are irresistibly attracted by the very anxieties we find most terrifying; we are drawn to them by a primitive excitement that arises from flirtation with danger. Moths and flames, mankind and death—there is little difference.

None of us seems psychologically able to cope with the thought of our own state of death, with the idea of a permanent unconsciousness in which there is neither void nor vacuum—in which there is simply nothing. It seems so different from the nothing that preceded life. As with every other looming terror and looming temptation, we seek ways to deny the power of death and the icy hold in which it grips human thought. Its constant closeness has always inspired traditional methods by which we consciously and unconsciously disguise its reality, such as folk tales, allegories, dreams, and even jokes. In recent generations, we have added something new: We have created the method of modern dying. Modern dying takes place in the modern hospital, where it can be hidden, cleansed of its organic blight, and finally packaged for modern burial. We can now deny the power not only of death but of nature itself. We hide our faces from its face, but still we spread our fingers just a bit, because there is something in us that cannot resist a peek.

We compose scenarios that we yearn to see enacted by our mortally ill beloved, and the performances are successful just often enough to sustain our expectations. Faith in the possibility of such a scenario has ever been a tradition of Western societies, which in centuries past valued a good death as the salvation of the soul and an uplifting experience for friends and family and celebrated it in the literature and pictorial representations of *ars moriendi*, the art of dying. Originally, *ars moriendi* was a religious and spiritual endeavor, described by the fifteenth-century printer William Caxton as "the craft for to deye for the helthe of mannes sowle." In time, it evolved into the concept of the beautiful death, truly the correct way to die. But *ars moriendi* is nowadays made difficult by the very fact of our attempts at concealing and sanitizing—and especially preventing—which result in the kinds of deathbed scenes that occur in such specialized hiding places as intensive care units, oncology research facilities, and emergency rooms. The good death has increasingly become a myth. Actually, it has always been for the most part a myth, but never nearly as much as today. The chief ingredient of the myth is the longed-for ideal of "death with dignity."

Not long ago, I saw in my clinical office a forty-three-year-old attorney on whom I had operated for an early-stage breast cancer three years before. Although she was free of disease and had every expectation of permanent cure, she seemed oddly upset that day. At the end of the visit, she asked if she might stay a bit longer, to talk. She then began to describe the recent death in another city of her mother, from the same disease of which she herself had almost certainly been cured. "My mother died in agony," she said, "and no matter how hard the doctors tried, they couldn't make things easy for her. It was nothing like the peaceful end I expected. I thought it would be spiritual, that we would talk about her life, about the two of us together. But it never happened—there was too much pain, too much Demerol." And then, in an outburst of tearful rage, she said, "Dr. Nuland, there was no dignity in my mother's death!"

My patient needed a great deal of reassurance that there had been nothing unusual about the way her mother died, that she had not done something wrong to prevent her mother from experiencing that "spiritual" death with dignity that she had anticipated. All of her efforts and expectations had been in vain, and now this very intelligent woman was in despair. I tried to make clear to her that the belief in the probability of death with dignity is our, and society's, attempt to deal with the reality of what is all too frequently a series of destructive events that involve by their very nature the disintegration of the dying person's humanity. I have not often seen much dignity in the process by which we die.

The quest to achieve true dignity fails when our bodies fail. Occasionally—very occasionally—unique circumstances of death will be granted to someone with a unique personality, and that lucky combination will make it happen, but such a confluence of fortune is uncommon, and, in any case, not to be expected by any but a very few people.

I have written this book to demythologize the process of dying. My intention is not to depict it as a horror-filled sequence of painful and disgusting degradations, but to present it in its biological and clinical reality, as seen by those who are witness to it and felt by those who experience it. Only by a frank discussion of the very details of dying can we best deal with those aspects that frighten us the most. It is by knowing the truth and being prepared for it that we rid ourselves of that fear of the terra incognita of death that leads to self-deception and disillusions.

There is a vast literature on death and dying. Virtually all of it is intended to help people cope with the emotional trauma involved in the process and its aftermath; the details of physical deterioration have for the most part not been much stressed. Only within the pages of professional journals are to be found descriptions of the actual processes by which various diseases drain us of vitality and take away our lives.

My career and my lifelong experience of death confirm John Webster's observation that there are indeed "ten thousand several doors for men to take their exits"; my wish is to help fulfill the prayer of the poet Rainer Maria Rilke: "Oh Lord, give each of us his own death." This book is about the doors, and the passageways that lead to them; I have tried to write it in such a way that insofar as circumstances allow, choices may be made that will give each of us his or her own death.

I have chosen six of the most common disease categories of our time, not only because they include the mortal illnesses that will take the great majority of us but for another reason as well: The six have characteristics that are representative of certain universal processes that we will all experience as we are dying. The stoppage of circulation, the inadequate transport of oxygen to tissues, the flickering out of brain function, the failure of organs, the destruction of vital centers—these are the weapons of every horseman of death. A familiarity with them will explain how we die of illnesses not specifically described in this book. Those I have chosen are not only our most common avenues to death, they are also the ones whose paving stones are trod by everyone, no matter the rarity of the final disease.

My mother died of colon cancer one week after my eleventh birthday, and that fact has shaped my life. All that I have become and much that I have not become, I trace directly or indirectly to her death. When I began writing this book, my brother had been dead just a little more than a year, also of colon cancer. In my professional and personal life, I have lived with the awareness of death's imminence for more than half a century, and labored in its constant presence for all but the first decade of that time. This is the book in which I will try to tell what I have learned.

New Haven
June 1993

Sherwin B. Nuland

AUTHOR'S NOTE

With the exception of Robert DeMatteis, the names of all patients and their families have been altered to preserve confidentiality. It should also be noted that "Dr. Mary Defoe," who appears in Chapter VIII, actually represents a composite of three young doctors at the Yale–New Haven Hospital.

HOW WE DIE

I

The Strangled Heart

EVERY LIFE IS different from any that has gone before it, and so is every death. The uniqueness of each of us extends even to the way we die. Though most people know that various diseases carry us to our final hours by various paths, only very few comprehend the fullness of that endless multitude of ways by which the final forces of the human spirit can separate themselves from the body. Every one of death's diverse appearances is as distinctive as that singular face we each show the world during the days of life. Every man will yield up the ghost in a manner that the heavens have never known before: every woman will go her final way in her own way.

The first time in my professional career that I saw death's remorseless eyes, they were fixed on a fifty-two-year-old man, lying in seeming comfort between the crisp sheets of a freshly made-up bed, in a private room at a large university teaching hospital. I had just begun my third year of medical school, and it was my unsettling lot to encounter death and my very first patient at the same hour.

James McCarty was a powerfully built construction executive whose business success had seduced him into patterns of living that we now know are suicidal. But the events of his illness took place almost forty years ago, when we understood a great deal less about the dangers of the good life—when smoking, red meat, and great slabs of bacon, butter, and belly were thought to be the risk-free rewards of achievement. He had let himself become flabby, and sedentary as well. Whereas he had once directed on-site the crews of his thriving construction company, he was now content to lead imperiously from behind a desk. McCarty delivered his pronouncements most of the day from a comfortable swivel chair that provided him an unobstructed view of the New Haven Green and the Quinnipiack Club, his favorite grillroom for midday executive gluttony.

The events of McCarty's hospitalization are easily recalled, because the startling staccato with which they burst forth instantly and permanently imprinted them in my mind. I have never forgotten what I saw, and did, that night.

McCarty arrived in the hospital's emergency room at about 8:00 p.m. on a hot and humid evening in early September, complaining of a constricting pressure behind his breastbone that seemed to radiate up into his throat and down his left arm. The pressure had begun an hour earlier, after his usual heavy dinner, a few Camels, and an upsetting phone call from the youngest of his three children, an indulged young woman who had just started her freshman year at a fashionable women's college.

The intern who saw McCarty in the emergency room noted that he looked ashen and sweaty and had an irregular pulse. In the ten minutes it took to wheel the electrocardiogram machine down the hall and connect it to the patient, he had begun to look better and his unsteady cardiac rhythm had reverted to normal. The electrocardiographic tracing nonetheless revealed that an infarction had occurred, meaning that a small area of the wall of the heart had been damaged. His condition seemed stable, and preparations were made to transfer him to a bed upstairs—there were no coronary intensive care units in the 1950s. His private physician came in to see McCarty and reassured himself that his patient was now comfortable and seemed to be out of danger.

McCarty reached the medical floor at 11:00 p.m., and I arrived with him. Not being on duty that evening, I had gone to the rush party that my student fraternity held to inveigle entering freshmen into joining. A glass of beer and a lot of conviviality had made me feel especially self-confident, and I decided to visit the care division to which I had been assigned only that morning, the first of my clinical rotations on the Internal Medicine service. Third-year medical students, who are just starting out in their initial experience with patients, tend to be eager to the point of zealousness, and I was no different than most. I came up to the division to trail after the intern, hoping to see an interesting emergency, and to make myself helpful in any way I could. If there was an imminent ward procedure, like a spinal tap or the placement of a chest tube, I wanted to be there to do it.

As I walked onto the division, the intern, Dave Bascom, took my arm as though he was relieved to see me. "Help me out, will you? Joe [the student on duty] and I are tied up down the hall with a bulbar polio that's going bad, and I need you to do the admission workup on this new coronary that's just going into 507—okay?"

Okay? Sure it was okay! It was more than okay; it was wonderful, exactly the reason I had returned to the division. Medical students of forty years ago were given much more autonomy than they are allowed today, and I knew that if I did the admission routines well, I would be granted plenty of work on the details of McCarty's recovery. I waited eagerly for a few minutes until one of the two nurses on duty had transferred my new patient comfortably from the gurney onto his bed. When she went scurrying down to the far end of the hall to help with the polio emergency, I slipped into McCarty's room and closed the door behind me. I didn't want to run the risk that Dave might come back and take over.

McCarty greeted me with a thin, forced smile, but he couldn't have found my presence reassuring. I have often wondered over the years what must have gone through the mind of that high-pressure boss of large, tough men when he saw my boyish (I was then twenty-two) face and heard me say that I had come to take his history and examine him. Whatever it was, he didn't get much chance to mull it over. As I sat down at his bedside, he suddenly threw his head back and bellowed out a wordless roar that seemed to rise up out of his throat from somewhere deep within his stricken heart. He hit his balled fists with startling force up against the front of his chest in a single synchronous thump, just as his face and neck, in the flash of an instant, turned swollen and purple. His eyes seemed to have pushed themselves forward in one bulging thrust, as though they were trying to leap out of his head. He took one immensely long, gurgling breath, and died.

I shouted out his name, and then I shouted for Dave, but I knew no one could hear me in the hectic polio room all the way down the corridor. I could have run down the hallway and tried to get help, but that would have meant the loss of precious seconds. My fingers felt for the carotid artery in McCarty's neck, but it was pulseless and still. For reasons I cannot explain to this day, I was strangely calm. I decided to act on my own. The possibility of getting into trouble for what I was about to attempt seemed a great deal less risky than letting a man die without at least trying to save him. There was no choice.

In those days, every room housing a coronary patient was supplied with a large muslin-wrapped package that contained a thoracotomy kit—a set of instruments with which the chest could be opened in the event of cardiac arrest. Closed-chest cardiopulmonary resuscitation, or CPR, had not yet been invented, and the standard technique in this situation was to attempt to massage the heart directly, by holding it in the hand and applying a long series of rhythmic squeezes.

I tore open the kit's sterile wrapping and grabbed the scalpel placed for ready access in a separate envelope on top. What I did next seemed absolutely automatic, even though I had never done it, or seen it done, before. With one surprisingly smooth sweep of my hand, I made a long incision starting just below the left nipple, from McCarty's breastbone around as far back as I could without moving him from his half-upright position. Only a little dark ooze leaked out of the arteries and veins I cut through, but no real flow of blood. Had I needed confirmation of the fact of death by cardiac arrest, this was it. Another long cut through the bloodless muscle, and I was in the chest cavity. I reached over to grab the double-armed steel instrument called a self-retaining retractor, slipped it in between the ribs, and turned its ratchet just far enough to allow my hand to squeeze inside and grasp what I expected to be McCarty's silent heart.

As I touched the fibrous sack called the pericardium, I realized that the heart contained within was wriggling. Under my fingertips could be felt an uncoordinated, irregular squirming that I recognized from its textbook description as the terminal condition called ventricular fibrillation, the agonal act of a heart that is becoming reconciled to its eternal rest. With unsterile bare hands, I grabbed a pair of scissors and cut the pericardium wide open. I took up Mr. McCarty's poor twitching heart as gently as I could and began the series of firm, steady, syncopated compressions that is called cardiac massage, intended to maintain a flow of blood to the brain until an electrical apparatus can be brought in to shock the fibrillating heart muscle back into good behavior.

I had read that the sensation imparted by a fibrillating heart is like holding in one's palm a wet, jellylike bagful of hyperactive worms, and that is exactly the way it was. I could tell by its rapidly decreasing resistance to the pressure of my squeezes that the heart was not filling with blood, and so my efforts to force something out of it were useless, especially since the lungs were not being oxygenated. But still I kept at it. And suddenly, something stupefying in its horror took place—the dead McCarty, whose soul was by that time totally departed, threw back his head once more and, staring upward at the ceiling with the glassy, unseeing gaze of open dead eyes, roared out to the distant heavens a dreadful rasping whoop that sounded like the hounds of hell were barking. Only later did I realize that what I had heard was McCarty's version of the death rattle, a sound made by spasm in the muscles of the voice box, caused by the increased acidity in the blood of a newly dead man. It was his way, it seemed, of telling me to desist—my efforts to bring him back to life could only be in vain.

Alone in that room with a corpse, I looked into its glazed eyes and saw something I should have noticed earlier—McCarty's pupils were fixed in the position of wide black dilatation that signifies brain death, and obviously would never respond to light again. I stepped back from the disordered carnage on that bed and only then realized that I was soaking wet. Sweat was pouring down my face, and my hands and my short white medical student's coat were drenched with the dark lifeless blood that had oozed out of McCarty's chest incision. I was crying, in great shaking sobs. I realized, too, that I had been shouting at McCarty, demanding that he live, screaming his name into his left ear as though he could hear me, and weeping all the time with the frustration and sorrow of my failure, and his.

The door swung open and Dave rushed into the room. With one glance he took in the entire scene, and understood it. My shoulders were heaving, and my weeping was by then out of control. He strode around to my side of the bed, and then, as if we were actors in an old World War II movie, he put his arm around my shoulders and said very quietly, "It's okay, buddy—it's okay. You did everything you could." He sat me down in that death-strewn place and began patiently, tenderly, to tell me all the clinical and biological events that made James McCarty's death inevitably beyond my control. But all I can remember of what he said, with that gentle softness in his voice, was: "Shep, now you know what it's like to be a doctor."

Poets, essayists, chroniclers, wags, and wise men write often about death but have rarely seen it. Physicians and nurses, who see it often, rarely write about it. Most people see it once or twice in a lifetime, in situations where they are too entangled in its emotional significance to retain dependable memories. Survivors of mass destruction quickly develop such powerful psychological defenses against the horror of what they have seen that nightmarish images distort the actual events to which they have been witness. There are few reliable accounts of the ways in which we die.

Nowadays, very few of us actually witness the deaths of those we love. Not many people die at home anymore, and those who do are usually the victims of drawn-out diseases or chronic degenerative conditions in which drugging and narcosis effectively hide the biological events that are occurring. Of the approximately 80 percent of Americans who die in a hospital, almost all are in large part concealed, or at least the details of the final approach to mortality are concealed, from those who have been closest to them in life.

An entire mythology has grown up around the process of dying. Like most mythologies, it is based on the inborn psychological need that all humankind shares. The mythologies of death are meant to combat fear on the one hand and its opposite—wishes—on the other. They are meant to serve us by disarming our terror about what the reality may be. While so many of us hope for a swift death or a death during sleep "so I won't suffer," we at the same time cling to an image of our final moments that combines grace with a sense of closure; we need to believe in a clear-minded process in which the summation of a life takes place—either that or a perfect lapse into agony-free unconsciousness.

The best-known artistic representation of the medical profession is Sir Luke Fildes' renowned 1891 painting entitled *The Doctor*. The scene is a simple fisherman's cottage on the coast of England, where a little girl lies quietly, seemingly unconscious, as death approaches. We see her grieving parents and the pensive, empathetic physician keeping his bedside vigil, powerless to weaken the tightening grip of mortality. When the artist was interviewed about the painting, he said, "To me, the subject will be more pathetic than any, terrible perhaps, but yet more beautiful."

Fildes clearly had to know better. Fourteen years earlier, he had seen his own son die of one of the infectious diseases that carried off so many children in those late-nineteenth-century years shortly before the dawn of modern medicine. We don't know what malady killed Phillip Fildes, but it could not have bestowed a peaceful ending on his young life. If it was diphtheria, he virtually choked to death; if scarlet fever, he probably had delirium and wild swings of high fever; if meningitis, he may have had convulsions and uncontrollable headaches. Perhaps the child in *The Doctor* has gone through such agonies and is now in the final peace of terminal coma—but whatever came in the hours prior to her "beautiful" passing must surely have been unendurable to the little girl and her parents. We rarely go gentle into that good night.

Francisco Goya, eight decades earlier, had been more honest—perhaps because he lived at a time when the face of death was everywhere. In his painting, variously called in English *Diphtheria* or *The Croup*, done in the style of the Spanish realist school and during a period of great realism in European life, we see a doctor holding a young patient's head steady with one hand on his neck while preparing to insert the fingers of his other hand down the boy's throat in order to tear out the diphtheritic membrane that will choke off his life if not removed. The original Spanish title of the picture, and of the disease, reveals the full force of Goya's directness, as well as that age's everyday familiarity with death. He called it *El Garrotillo*, for the strangulation by which it kills its victims. The days of such confrontations with the reality of death are long since over, at least in the West.

Having chosen, for whatever psyche-shrouded reason, the word *confrontations*, I need to pause; I need to consider whether I, too, even after almost forty years of James McCartys, do not from time to time still fall into stride with the prevailing temperament of our times, when death is regarded as the final and perhaps the ultimate challenge of any person's life—a pitched battle that must be won. In that view, death is a grim adversary to be overcome, whether with the dramatic armaments of high-tech biomedicine or by a conscious acquiescence to its power, an acquiescence that evokes the serene style for which present usage has invented a term: "Death with dignity" is our society's expression of the universal yearning to achieve a graceful triumph over the stark and often repugnant finality of life's last sputterings.

But the fact is, death is not a confrontation. It is simply an event in the sequence of nature's ongoing rhythms. Not death but disease is the real enemy, disease the malign force that requires confrontation. Death is the surcease that comes when the exhausting battle has been lost. Even the confrontation with disease should be approached with the realization that many of the sicknesses of our species are simply conveyances for the inexorable journey by which each of us is returned to the same state of physical, and perhaps spiritual, nonexistence from which we emerged at conception. Every triumph over some major pathology, no matter how ringing the victory, is only a reprieve from the inevitable end.

Medical science has conferred on humanity the benison of separating those pathological processes that are reversible from those that are not, constantly adding to the means by which the balance shifts ever in favor of sustained life. But modern biomedicine has also contributed to the misguided fancy by which each of us denies the certain advent of our own individual mortality. The claims of too many laboratory-based doctors to the contrary, medicine will always remain, as the ancient Greeks first dubbed it, an Art. One of the most severe demands that its artistry makes of the physician is that he or she become familiar with the poorly delineated boundary zones between categories of treatment whose chances of success may be classified as certain, probable, possible, or unreasonable. Those unchartable spaces between the probable and everything beyond it are where the thoughtful physician must often wander, with only the accumulated judgment of a life's experiences to guide the wisdom that must be shared with those who are sick.

At the time that James McCarty's life came to its abrupt end, the outcome of his heart's misbehavior was inescapable. Although a great deal was already understood about heart disease in the early 1950s, the available therapies for it were few and too often inadequate. Today, a patient with McCarty's specific problem may expect to leave the hospital not only alive but with a heart so much improved that years may have been added to his life. So much have the laboratory-based doctors accomplished that one of the approximately 80 percent who survive a first attack has good reason to think of a cardiac seizure as the shiniest silver lining of his life, because it has exposed a condition that might soon have killed him had it not been discovered while still eminently treatable.

Indeed, the balance has shifted so much that the effectiveness of treatment for cardiac disease is far more often on the good side of probable. That should not, however, be taken to mean that the once imperilled heart is now an immortal heart. Although the great majority of cardiac patients today survive their first episode, well over half a million Americans still die every year of some form of McCarty's sickness. Another 4.5 million are newly diagnosed as being afflicted with it. Eighty percent of people whose heart disease eventually kills them are victims of this particular form of it: Ischemic heart disease (or coronary artery disease, or coronary heart disease, as it is variously called) is the leading cause of death in the industrialized nations of the world.

James McCarty's heart died because it was not getting enough oxygen; it was not getting enough oxygen because it was not getting enough hemoglobin, the blood-borne protein whose function is to carry the oxygen; it was not getting enough hemoglobin because it was not getting enough blood; it was not getting enough blood because the heart's nourishing vessels, the coronary arteries, were hardened and narrowed by a process called arteriosclerosis (literally, hardening of the arteries). The arteriosclerosis had occurred because of a combination of McCarty's sybaritic diet, his cigarette smoking, his lack of exercise, an element of high blood pressure, and a certain degree of inherited predisposition. Very likely, the phone call from his pampered daughter had the same spasm-inducing effect on his severely narrowed coronary arteries as it did on his angrily clenched fists. That bit of acute tightening was probably just enough to rupture or crack one of the deposits of arteriosclerosis, called plaques, in the lining of a main coronary artery. Once this occurred, the disrupted plaque served as a focus on which fresh blood-clot formed, making the obstruction complete and choking off the already-compromised flow. This final stoppage caused so-called "ischemia" (pronounced *iskeemeeya*), or blood lack, thereby acutely starving a large-enough piece of McCarty's heart muscle, or myocardium, to disrupt its normal rhythm into the chaotic squirming of ventricular fibrillation.

It is quite possible that none of McCarty's heart muscle was actually killed by its acute blood lack. Ischemia alone may cause ventricular fibrillation, especially in a heart already injured by a previous attack And so may the adrenalinelike compounds produced by the body at times of stress. Whatever the cause, the electrical communication system upon which James McCarty's heart depended for its regularity and coordination broke down, and so did McCarty's life.

Like so many other medical terms, *ischemia* is a word with an interesting history and colorful associations. It will recur again and again in the telling of the stories in this long narrative of death, because it is so ubiquitous—and so insidious—a driving force toward the quenching of life's energies. Though starvation of the heart may offer the most dramatic example of its lurking dangers, the process of choking off oxygen and nutrition is the common denominator in a wide variety of mortal illnesses.

The concept of ischemia and the word itself were introduced in the middle of the nineteenth century by a brashly brilliant little Pomeranian (the word, when applied to dogs, evokes a tiny and intensely spirited bundle of scrappy exuberance, which seems appropriate for the man being described) who began his multifaceted career as a kind of *enfant terrible* of research, and ended it sixty years later universally recognized by the sobriquet "the Pope of German Medicine." No single individual has ever contributed more to the understanding of the ways in which disease wreaks its havoc on human organs and cells than did Rudolf Virchow (1821–1902).

Virchow, a professor of pathology at the University of Berlin for almost fifty years, produced more than two thousand books and articles, not only on medicine but on anthropology and German politics as well. So liberal a member of the Reichstag was he that the autocratic Otto von Bismarck once challenged him to a duel. Being given the choice of weapons, Virchow ridiculed the upcoming encounter out of existence before it took place—by insisting that it be fought with scalpels.

Among Rudolf Virchow's many research interests was his fascination with the ways in which disease affects arteries, veins, and their contained blood constituents. He elucidated the principles of embolism, thrombosis, and leukemia and invented the words to describe them. Seeking a term to designate the mechanism by which cells and tissues are deprived of their blood supply, Virchow seized (this word is chosen advisedly) upon the Greek *ischano*—"I hold in check," or "I quench"—derived from the Indo-European root *segh*, which refers to "seizing" or "holding" or "causing to pause." By combining it with *aima*, or "blood," the Greeks had created the word *ischaimos*, to signify a holding in check of the flow of blood. *Ischemia* was chosen by Virchow to designate the consequences of diminishing or totally stopping blood flow to some structure of the body, whether as small as a cell or as large as a leg or a section of heart muscle.

Diminishing is a relative term, however. When an organ's activity increases, its oxygen requirements go up, and so does its need for blood. If narrowed arteries cannot widen to accommodate this need, or if for some reason they go into tight spasm that further restricts flow, the organ's demands are not met, and it rapidly becomes ischemic. In pain and anger, the heart screams out a warning, and continues to do so until its shrieking exhortations for more blood are met, usually by the natural stratagem of the victim, who—alarmed by the distress within his chest—slows or stops the activity that is tormenting his cardiac muscle.

A ready example of this process is the suddenly overworked calf muscle of a weekend athlete who returns to jogging each year when the weather warms up in April. The discrepancy between the amount of blood required by his out-of-condition muscle and the amount that is able to force its way through his out-of-condition arteries may result in ischemia. The calf does not get enough oxygen and it cries out in an agonizing seizure, to warn the athlete manqué to stop his exertions before a clump of muscle cells are starved to death, the process known as infarction. The shriek of pain in the overtaxed calf is called a cramp or a charley horse. When it originates in the heart muscle, we use the much more elegant term *angina pectoris*. Angina pectoris is nothing else than a charley horse of the heart. If it lasts long enough, its victim sustains a myocardial infarction.

Angina pectoris is a Latin phrase which translates literally as "a choking" or "throttling" (*angina*) "of the chest" (*pectoris*, the genitive case of *pectus*, "chest"). It is to another medical philologist, the remarkable eighteenth-century English physician William Heberden (1710–1801), that we owe not only the term but also one of the finest descriptions of the symptoms associated with it. In a 1768 discussion of the various forms of chest pain, he wrote:

> But there is a disorder of the breast marked with strong and peculiar symptoms, considerable for the kind of danger belonging to it, and not extremely rare, which deserves to be mentioned more at length. The seat of it, and sense of strangling and anxiety with which it is attended, may make it not improperly be called angina pectoris.
>
> They who are afflicted with it, are seized while they are walking, (more especially if it be up hill, and soon after eating) with a painful and most disagreeable sensation in the breast, which seems as if it would extinguish life, if it were to increase or to continue; but the moment they stand still, all this uneasiness vanishes.

Heberden had seen enough patients—"nearly a hundred under this disorder"—to be able to study its incidence and progress:

> Males are most liable to this disease, especially such as have past their fiftieth year.
>
> After it has continued a year or more, it will not cease so instantaneously upon standing still; and it will come on not only when the persons are walking, but when they are lying down, especially if they lie on the left side, and oblige them to rise up out of their beds. In some inveterate cases it has been brought on by the motion of a horse, or a carriage, and even by swallowing, coughing, going to stool, or speaking, or by any disturbance of mind.

Heberden was struck by the unremitting progression of the disease: "For if no accident intervene, but the disease go on to its height, the patients all suddenly fall down, and perish almost immediately."

James McCarty never had the luxury of a succession of bouts of angina pectoris; he succumbed to his very first experience of cardiac ischemia. His brain died because the fibrillating and finally stilled heart could no longer pump blood to it. The ischemic brain was followed gradually into lifelessness by every other tissue in his body.

A few years ago, I met a man who was miraculously resuscitated from such an apparent sudden cardiac death. Irv Lipsiner is a tall, broad-shouldered stockbroker who has been an avid athlete all his life. Although he requires insulin for long-standing diabetes, the disease has had no physical effects on his vigorous good health, or so it would appear at first glance. But he did have a small heart attack when he was forty-seven years old, which is exactly the age at which his father died from the same cause. That episode left his heart muscle with only minimal damage, and he continued his active life without restriction.

Late on a Saturday afternoon in 1985, when he was fifty-eight years old, Lipsiner was beginning his third hour of tennis at the Yale indoor courts when two of his partners left, necessitating a switch from doubles to singles. The practice rally was just beginning when, without warning or premonitory pain, he slumped to the floor unconscious. Two physicians, by luck playing on an adjacent court, rushed to his aid and found him glassy-eyed, unresponsive, and not breathing. There was no heartbeat. Assuming correctly that he was in ventricular fibrillation, they immediately began cardiopulmonary resuscitation, continuing it for what seemed to them an interminable time, until the ambulance arrived. By then, Lipsiner had begun to respond, even resuming a spontaneous regular heartbeat as his airway was intubated and he was placed in the ambulance. Soon, he was wide awake in the Yale–New Haven Hospital emergency room and wondering, as he put it, "what the fuss was all about."

In two weeks, Lipsiner was out of the hospital, fully recovered from his episode of ventricular fibrillation. I met him some years later, on the horse farm where he lives. Every day, he takes time out from work to go riding or play tennis, usually singles. Here is

Irv Lipsiner describing what it felt like to drop dead on a tennis court:

> The only thing I can recall is just—not hurting, but just collapsing. And then the lights went out, as if you're in a little room and you flip the switch. The only thing different from that was that it was in slow motion. In other words, it didn't go out like *that* [here he snapped his fingers]. It went out like this [he made a lazy downward circle with his hand, like an airplane turning gently in descent toward a landing], gradually and almost in a spiral, like—[he hesitated briefly in thought, then pursed his lips and blew his breath out in a slow diminuendo]—this. The change from light to dark was very evident, but the speed with which it happened was—well, gradual.
>
> I was aware that I'd collapsed. I felt like somebody took the life out of me. It felt like—I'm thinking of a scene—I had a dog that was hit by a car, and when I looked at that dog on the ground—he was dead already—he just looked like the same dog, only shrunk. You know, shrunk—uniformly. That's how I felt. I felt like—[he made a sound like air going out of a balloon] "Pffft."

Lipsiner's light went out precisely the way it did because the circulation to his brain had been suddenly shut off. As the oxygen in the organ's now-stagnant blood was steadily used up, the brain began to fail—sight and consciousness were turned down as though by the gradual twist of a dial rather than the suddenness of a switch. That was Irv Lipsiner's slow-motion spiral into oblivion, and almost death. The mouth-to-mouth breathing and chest massage of the cardiopulmonary resuscitation forced air into his lungs and drove blood to his vital organs until his heart decided, for reasons of its own, to resume its responsibilities. Like most sudden cardiac deaths in nonhospitalized people, Irv Lipsiner's episode was caused by ventricular fibrillation.

Lipsiner felt no ischemic pain. The probable cause of his fibrillation was some transient chemical stimulation of a supersensitive area left on his heart muscle by the attack of 1974. As to why the fibrillation occurred when it did, there is no way to be certain; but a quite plausible guess is that it was related to the stress of too much tennis on that Saturday afternoon, which could have caused the release into his circulation of extra adrenaline, and this in turn may have made a coronary artery go into spasm and set off the irregular rhythm. Such are the occasional vagaries of ischemic heart disease that Lipsiner was left with no new damage to his heart, although he never again played more than two consecutive hours of tennis.

The fact that Lipsiner experienced no cardiac charley horse before he began fibrillating makes this particular case of heart seizure somewhat unusual—the majority of people who drop dead probably do feel ischemic pain of the characteristic sort. Like its equivalent in the calf, the onset of ischemic cardiac pain is sudden and severe. It has been most commonly described by its sufferers as constricting, or viselike. Sometimes it manifests itself as a crushing pressure, like an intolerable blunt weight forcing itself against the front of the chest and radiating down the left arm or up into the neck and jaw. The sensation is frightening even to those who have experienced it often, because each time it recurs it is accompanied by awareness of the possibility (and quite a realistic awareness it is) of impending death. The sufferer is likely to break out into a cold sweat, feel nauseated, or even vomit. There is often shortness of breath. If the ischemia does not let up within approximately ten minutes, the oxygen deficiency may become irreversible, and some of the deprived cardiac muscle will go on to die, the process called myocardial infarction. If that happens, or if the oxygen lack is sufficient to scramble the heart's conduction system, some 20 percent of the afflicted will perish in the throes of such an episode before reaching an emergency room. That figure drops by at least half if transportation to a hospital is possible within the period cardiologists call "the golden hour."

Eventually, about 50 to 60 percent of people with ischemic heart disease will die within an hour of one of their attacks, whether the first or a later one. Since 1.5 million Americans suffer a myocardial infarction each year (70 percent of which occur in the home), it is not difficult to understand why coronary heart disease is America's biggest killer, as it is in every industrialized country of the world. Almost all of those who survive every infarction will eventually be claimed by the gradual weakening of the heart's ability to pump.

When all natural causes are taken into account, approximately

20 to 25 percent of Americans die suddenly, defined as unexpected death within a few hours of onset of symptoms in persons neither hospitalized nor homebound. And of these deaths, 80 to 90 percent are cardiac in origin, the remaining segment being due to diseases of the lungs, central nervous system, or the vessel into which the left ventricle pumps its blood, the aorta. When the death is not only sudden but instantaneous, there are only a few that are not the result of ischemic heart disease.

The victims of ischemic heart disease are betrayed by their eating and their smoking and their inattention to such simple housekeeping chores as exercise and the maintenance of normal blood pressure. Sometimes pedigree alone gives them away, in the form of family history or diabetes; sometimes it is that driving impetuosity and aggressiveness that today's cardiologists call the Type A personality. In a way, the person whose heart muscle will be anguished by angina is very like the overly ambitious schoolchild who throws a hand aggressively into the air when the teacher looks for volunteers—"Choose me, choose me; I can do it better than anyone else!" He is easy to identify, and death will single him out. There is little randomness in the choices made by cardiac ischemia.

Long before we knew about the lurking perils of cholesterol, cigarettes, diabetes, and hypertension, the medical world was beginning to recognize specific characteristics in those persons who seemed destined for cardiac death. William Osler, the author of America's first great textbook of medicine in 1892, might have been describing James McCarty when he wrote, "It is not the delicate neurotic person who is prone to angina, but the robust, the vigorous in mind and body, the keen and ambitious man, the indicator of whose engines is always at 'full speed ahead.' " By their speedometers shall ye know them.

Despite all medical advances, there are still plenty of people who die with their first heart attack. Like lucky Lipsiner, most of them do not actually suffer death of cardiac muscle but are victimized by a rhythm suddenly made disorderly by the effect of ischemia (or sometimes local chemical changes) on an electrical conduction system already sensitized by a previous injury, whether it was recognized or not. But the usual way in which people succumb to ischemic heart disease these days is not the way of Lipsiner or McCarty. Decline is most often gradual, with plenty of warnings and much successful treatment before the final summons. The killing off of increments of heart muscle takes place over a period of months or years, until that besieged and enervated pump simply fails. It then gives up, for lack of strength or because the command system that controls its electrical coordination can no longer recover from yet another breach of its authority. Those laboratory doctors who are convinced that medicine is a science have accomplished so much that those bedside doctors who know it is an art can often, by careful timing and skillful choice of what is now available to them, provide victims of heart disease with long periods of improvement and stable health.

The fact remains, however, that each day fifteen hundred Americans will die of cardiac ischemia, whether its course has been sudden or gradual. Although preventive measures and modern methods of treatment have been reducing the figure steadily since the mid-1960s, it is virtually impossible for any slope of decline to change the picture for the vast majority of those who carry the diagnosis today or in whom it will be made in the next decade. This unforgiving sickness, like so many other causes of death, is a progressive continuum whose ultimate role in our planet's ecology is the quenching of human life.

In order to make clear the sequence of events that leads to the gradual loss of a heart's ability to pump effectively, it is first necessary to review some of the wondrous qualities that enable it to perform with such extraordinary precision when it is healthy. This will be the subject of the first pages of the chapter that follows.

II

A Valentine—and How It Fails

AS EVERY CHILD knows, the heart is shaped very much like a valentine. It is made almost entirely of muscle, called myocardium, wrapped around a large central space that is subdivided into four chambers: A vertical front-to-back wall of tissue, called the septum, separates the large space into right and left portions, and a transverse sheet at right angles to the septum divides each of those portions into upper and lower parts, making four in all. Because they have a certain degree of independence from one another, the portions on either side of the vertical septum are often called the right and left heart. On each side, the transverse sheet separating top from bottom is perforated by a central opening fitted with a one-way valve that allows blood to pass easily from the upper chamber (called the atrium) down into the lower chamber (the ventricle). In a healthy heart, the valves close tightly when the ventricle is filled, to prevent blood from regurgitating back up into the atrium. The atria are primarily receiving chambers, and the ventricles are pumping chambers. Consequently, the portion of the cardiac muscle around the upper part of the heart does not have to be as thick as that of the more powerful ventricles below them.

In a sense then, we have not one heart but two, attached side by side to each other by the septum; each has an upper chamber to receive and a lower one to pump. The two hearts have quite different jobs to do: The function of the right is to receive "used" blood returning from the tissues and drive it the short distance through the lungs, where it will be freshly aerated with oxygen; the left heart, in turn, receives the oxygen-rich blood returning

Exterior of a normal adult heart showing the coronary arteries

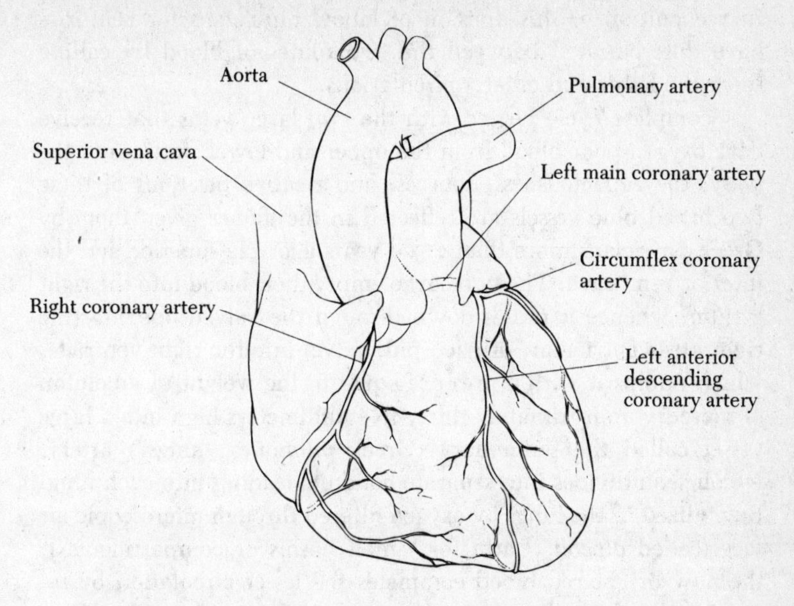

Diagrammatic section cut through a normal heart with arrows indicating the blood flow

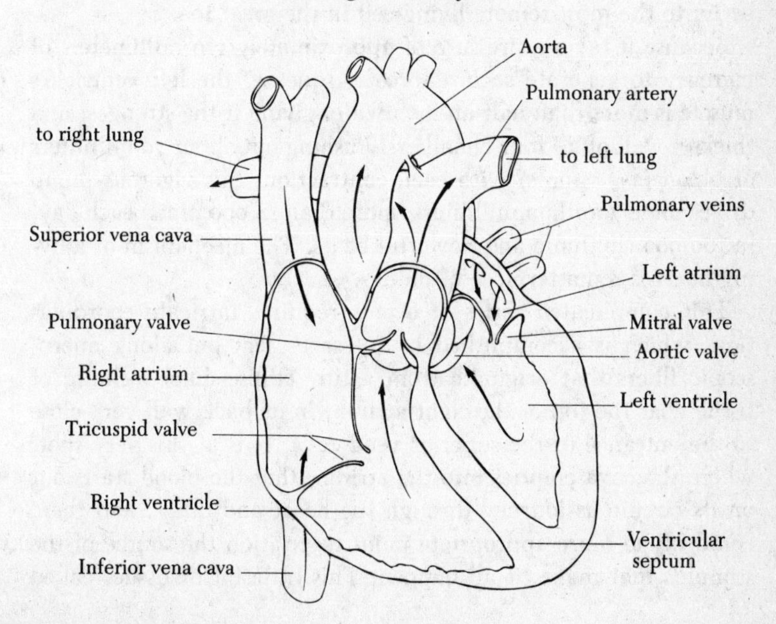

from the lungs and forcefully pumps it out to the rest of the body. In recognition of this division of labor, physicians for centuries have differentiated between the two routes of blood by calling them the lesser and greater circulations.

A complete cycle begins with the two large veins that receive dark oxygen-poor blood from the upper and lower portions of the body; the capaciousness, sources, and relative positions of these two broad blue vessels are reflected in the names given them by Greek physicians more than 2,500 years ago: the superior and the inferior vena cava. The two cavae empty their blood into the right atrium, whence it passes down through the valved opening (the right atrioventricular, or tricuspid, valve) into the right ventricle, which pumps it with a pressure equal to the weight of a column of mercury approximately thirty-five millimeters high into a large vessel called the pulmonary (Greek *pulmone*, "lungs") artery, which soon divides into separate conduits leading into each lung. Revitalized in the lungs by oxygen filtered through microscopic air sacs (called *alveoli*, Latin for "small basins or compartments"), the now bright red blood completes the lesser circulation by returning via the pulmonary veins to the left atrium, to be channeled down into the ventricle and thence driven throughout the body, to the most remote living cell in the great toe.

Because it takes a pressure of approximately 120 millimeters of mercury to generate such a forceful squeeze, the left ventricle's muscle is more than half an inch wide, giving it the strongest and thickest wall of all four chambers. Pushing out about 70 milliliters of blood (2⅓ ounces) with each contraction, this vigorous pump drives some 7 million milliliters (more than 14,000 pints) each day, in 100,000 rhythmic and powerful beats. The mechanism of a living heart is a masterpiece of nature.

This complicated series of events requires intricate coordination, which is accomplished by messages sent out along microscopic fibers that originate from a tiny ellipse-shaped clump of tissue near the top of the right atrium, in its back wall very close to the entrance of the superior vena cava. It is at this very spot, where the cava empties into the atrium, that the blood starts out on its circuitous journey through the heart and lungs, and there could be no more appropriate point to position the source of the stimulus that makes it all happen. This little bit of tissue, called the sinoatrial (or SA) node, is a pacemaker that drives the coordinated beating of the heart. A bundle of fibers carries the SA node's messages to a relay station lying between the atria and ventricles (and therefore called the atrioventricular, or AV, node), and from there they are transmitted to the muscle of the ventricles via an arborizing network of fibers called the bundle of His, named for its discoverer, a nineteenth-century Swiss anatomist who spent most of his career at the University of Leipzig.

The SA node is the heart's personal internal generator; nerves from outside may affect the rate of beating, but it is the conduction of electricity from the SA node that determines the wondrous regularity of its faultless rhythm. Awestruck anew each time they viewed the proud independence of an exposed animal heart, wise men of ancient civilizations proclaimed that this supernal mechanism of boldly autonomous flesh must be the dwelling place of the soul.

The blood within the heart's chambers is only passing through; it does not stop to nourish the muscular valentine whose syncopated strokes are busily squeezing it along its way through the circulation. For the sustenance required for its forceful labors, the heart muscle, or myocardium, is supplied by a group of separate and distinct vessels, which, because they originate in encircling arteries that wind around the heart like a crown, are called coronary. Branches of the main coronary arteries descend toward the valentine's tip, giving off twiglike branchlets that bring bright red oxygen-rich blood to the rhythmically heaving myocardium. In health, these coronary arteries are the friends of the heart; when they are diseased, they betray it at its most needful moments.

So commonly do the coronary arteries betray the heart whose muscle they are meant to sustain, that their treachery is the cause of at least half of all deaths in the United States. These "now I love you, now I don't" vessels are gentler to the gentle sex than they are to those who have more commonly gone out to hunt and fish—not only is infarction less common in women, it tends also to come later in life. The average age of the first infarction of women is in the mid-sixties, but men are more likely to have that terrifying experience ten years earlier. Although the coronary arteries have by that age reached the critical degree of narrowing necessary to threaten the viability of heart muscle, the process begins when its victims are much younger. An oft-quoted study of soldiers killed in the Korean War revealed that some three-quarters of these young men already had some arteriosclerosis in their coronary vessels. Varying degrees of it can be found in virtually every American adult, having begun with adolescence and increasing with age.

The obstructing material takes the form of yellowish white clumps called plaques, which are densely adherent to the inner lining of the artery and protrude into its central channel. The plaques are made up of cells and connective tissue, with a central core composed of debris and a common variety of fatty material called lipid, from the Greek *lipos*, meaning "fat" or "oil." Because so much of its structure is lipid, a plaque is called an atheroma, from the Greek *athere*, meaning "gruel" or "porridge," and *oma*, signifying a growth or tumor. The process of atheroma formation being by far the most common cause of arteriosclerosis, it is usually referred to as atherosclerosis, hardening by atheroma.

As an atheroma progresses, it becomes larger and tends to coalesce with neighboring plaques at the same time that it is absorbing calcium from the bloodstream. The result is the gradual accumulation of an extensive mass of crusted atheroma that lines a vessel for a considerable distance, making it increasingly gritty, hard, and narrowed. An atherosclerotic artery has been compared to an old length of much-used, poorly maintained pipe whose inner diameter is lined with thick, irregular deposits of rust and embedded sediment.

Even before the cause of angina pectoris and infarction was understood to be a narrowing of the coronary arteries, a few physicians were beginning to make observations about the hearts of those people who died of the process. The same Edward Jenner who introduced smallpox vaccination in 1798 was an inveterate student of disease who made a custom of following to the autopsy table as many of his deceased patients as possible. In those days, doctors performed their own postmortem examinations. As a result of his dissections, Jenner began to suspect that the narrowing he discovered in the death-room coronary arteries was directly related to the anginal symptoms he had elicited from patients during life. In a letter to a colleague, he wrote of a recent experience dissecting a heart during such an autopsy:

> My knife struck something so hard and gritty as to notch it. I well remember looking up at the ceiling, which was old and crumbling, conceiving that some plaster had fallen down. But on further scrutiny the real cause appeared: the coronaries were become bony canals.

In spite of Jenner's observations and a gradual increase in understanding the way in which coronary obstruction injures the heart, it took until 1878 before a physician was able to diagnose a myocardial infarction correctly. Dr. Adam Hammer of St. Louis, a German refugee from the repression following the unsuccessful revolutions of 1848, sent off to a medical journal in Vienna his case report, entitled "*Ein Fall von thrombotischem Verschlusse einer der Kranzarterien des Herzens,*" "A Case of Thrombotic Occlusion of One of the Coronary Arteries of the Heart." (Here an interesting twist of language presents itself: The German term for coronary artery is *Kranzarterie*, a *Kranz* being a wreath or a crown of flowers, which thus bestows an entirely new and quite poetic significance on the valentine image.) Hammer had been called in consultation to see a suddenly stricken thirty-four-year-old man who was in such a rapidly worsening state of collapse that death was imminent. Although physicians knew the mechanism of myocardial ischemia, the diagnosis of infarction caused by it had never been made, or even thought of. As he watched helplessly while his patient died, Hammer suggested to his colleague that a completely occluded coronary artery had caused death of heart muscle, and he decided that an autopsy was mandatory to prove his novel theory. It was no easy matter to obtain permission from the grief-stricken family, but the experienced Hammer overcame their objections by the timely application of that perennial solvent of reluctance, a handful of dollars. As he put it so frankly in his journal article: "In the face of this universal remedy, even the most subtle misgivings, including the religious ones, eventually yield." Hammer's persistence was rewarded by finding a pale yellow-brown myocardium (its color signifying infarction) and a completely occluded coronary artery, confirming his insight.

During the following decades, the principles of ischemic heart disease and infarction became gradually established. With the invention of the electrocardiogram in 1903, physicians were able to trace the messages carried by the heart's conduction system of fibers, and they soon learned to interpret the tracings made by electrical changes taking place when the heart muscle is endangered by a decreased blood supply. Other diagnostic techniques were discovered apace, including the fact that injured myocardium releases certain chemicals or enzymes whose identifiable presence in the blood aids in detecting infarction.

An individual infarction involves that part of the muscle wall supplied by the particular coronary artery that is occluded, a part that most commonly measures two or three square inches in surface area. The specific culprit almost half the time is the left anterior descending coronary artery, a vessel that passes down the front surface of the left heart toward its tip, tapering as it gives off subdivisions that enter the myocardium. The frequent involvement of this artery means that approximately half of infarctions involve the front wall of the left ventricle. Its back wall is supplied by the right coronary artery, which accounts for 30 to 40 percent of occlusions; the lateral wall is supplied by the left circumflex coronary artery, which contributes 15 to 20 percent.

The left ventricle, the most powerful part of the cardiac pump and the source of the muscular strength that nourishes every organ and tissue of the body, is injured in virtually every heart attack—each cigarette, each pat of butter, each slice of meat, and each increment of hypertension make the coronary arteries stiffen their resistance to the flow of blood.

When a coronary artery suddenly completes the process of occlusion, a period of acute oxygen deprivation ensues. If the oxygen lack is of such duration and severity that the stunned and instantly bloodless muscle cells cannot recover, the pain of angina is succeeded by infarction: The affected muscle tissue of the heart goes from the extreme pallor of ischemia to frank death. If the area of death is small enough and has not killed the patient by causing ventricular fibrillation or some equally serious abnormality of rhythm, the involved muscle, now puffy and swollen, will be able to maintain a tenuous hold on existence until, with the process of gradual healing, it is replaced by scar tissue. The area of such tissue is incapable of participating in the forceful thrusting of the rest of the myocardium. Each time a person recovers from a heart attack of any size, he has lost a little more muscle to the increasing area of scar tissue, and the power of his ventricle becomes just a bit less.

As atherosclerosis progresses, the ventricle may gradually weaken even when there is no frank heart attack. Coronary occlusions in smaller branches of the main vessels may give no signals to announce themselves, but they continue nevertheless to diminish the force of cardiac contraction. Eventually, the heart begins to fail. It is the chronic disease of heart failure, and not the sudden end of the James McCartys, that carries off approximately 40 percent of the victims of coronary artery disease.

Differing combinations of instigating circumstances and tissue damage determine the type and degree of danger in which each individual heart finds itself at any particular stage of its decline. One or another factor may predominate at a given point: Sometimes it is the susceptibility to spasm or thrombosis of the partially occluded coronary arteries; sometimes it is the sick cardiac muscle whose damaged communication system is so confused and hyperexcitable that it fibrillates with minimal stimulus; sometimes it is the communication system itself, which becomes sluggishly loath to transmit signals, so that it falters, slows, or even allows the heart to stop altogether; sometimes it is a ventricle too scarred and weakened to eject a sufficient fraction of the blood that has poured into it from its atrium.

When the 20 percent of cardiac patients who die in a McCarty-like first attack are added to those who expire suddenly after weeks or years of worsening disease, the total figure for sudden deaths amounts to some 50 to 60 percent of people who have ischemic heart disease. The remainder die slowly and uncomfortably of one of the variations of what is called chronic congestive heart failure. Although (or perhaps because) the death rate from heart attacks has decreased by approximately 30 percent in the past two or three decades, mortality due to congestive heart failure has gone up by one-third.

Chronic congestive failure is the direct result of the scarred and weakened myocardium's inability to contract with enough force to push out the necessary volume of blood with each stroke. When the blood that has already entered the heart cannot be efficiently pumped forward into the greater and lesser circulations, some of it backs up into the veins that are returning it, causing backpressure in the lungs and other organs from which it is coming. The result of this congestion is to drive some of the blood's fluid component through the leaking walls of the smallest vessels, resulting in swelling, or edema, of tissue. Structures like the kidney and liver are thus prevented from performing efficiently, a state of affairs made even worse by the fact that the left ventricle's weakened pump drives less of the newly oxygenated blood it receives, decreasing even the nourishment of the already-swollen organs. In this way, the general slowing down of the circulation is accompanied by a decrease in the flow of blood in and out of tissues.

The backpressure of inadequately ejected blood causes the heart's chambers to balloon out and remain dilated. The ventricular muscle thickens in an attempt to compensate for its own weakness. Thus the heart becomes enlarged and appears more formidable, but it is now only a thing of blustering braggadocio. Huffing and puffing, it speeds up the rate of its beat, trying to put out more blood. Before long, it finds itself in the ever-worsening plight of having, Alice-like, to run faster just to keep up. The exertions of the distended, thickened heart require more oxygen than the narrowed coronary arteries can bring it, and the faltering myocardium may be damaged further, or perhaps new abnormalities of rhythm will appear. Some of these abnormalities are lethal—ventricular fibrillation and similar disturbances of rhythm kill almost half of the patients in heart failure. So, no matter how boastful its bombast, the failing heart continues to fail, in a kind of vicious circle of trying to disguise its own inadequacies by straining to compensate for them. As a cardiologist colleague has put it, "Heart failure begets heart failure." The proprietor of that heart is beginning to die.

The afflicted patient becomes increasingly short of breath with even minimal exertion, since neither the heart nor the lungs can respond to the increase in the work demanded of them. Some sufferers have difficulty lying down for more than a short period of time, because they need the upright position and gravity's help to drain excess fluid from their lungs. I have known many patients for whom sleep became impossible unless their head and shoulders were elevated on several pillows, and even then they were subject to paroxysms of frightening breathlessness during the night. Patients in heart failure suffer also from chronic fatigue and listlessness, owing to a combination of the added effort of breathing and the poor tissue nutrition caused by low cardiac output.

The elevated pressure that is transmitted from the venae cavae back into the body's veins causes the feet and ankles to swell, but when patients are bedridden, gravity forces the fluid to collect in the tissues of the lower back and thighs. Although rare today, it was not uncommon in my medical school years to come upon a patient sitting upright in bed, belly and legs swollen with fluid, throwing himself into almost convulsive heavings of shoulders and gaping mouth while struggling fiercely for each individual gasping breath as if it were his last chance to save his own life. In the wide-open mouths of these combatants in losing campaigns against imminent mortality, one could usually detect the blueness of deoxygenated lips and tongues, parchment-dry even though the dying patients were drowning. Doctors feared to do anything that might worsen the already intolerable eye-bulging anxiety of a man being submerged in his own waterlogged tissues, hearing only the horrible wheeze and gurgling of his own death agony. In those days, we had little to offer a terminal sufferer except sedation, with the full and merciful knowledge that every bit of relief brought the end closer.

Although nowadays less common, such scenes are sometimes still enacted. A professor of cardiology recently wrote me: "There are many patients with terminal, intractable congestive heart failure whose final hours—or days—of life are made uncomfortable and even miserable by their drowning, while physicians can only watch helplessly, and use morphine for sedation. It isn't a pleasant exit." Not only the heart itself but the long-range damage inflicted by soggy, anemic tissues has plenty of other ways to kill. Eventually, the abused organs themselves fail. When the kidneys or liver are gone, so, too, is life. Kidney failure, or uremia, is an exit for some cardiac patients and so, on occasion, is inadequacy of liver function, frequently signaled by the appearance of jaundice.

Not only does the heart fool itself into overactivity, it may also fool the organs that might be able to help it out of its troubles. The kidney should be able to filter enough extra salt and water out of the blood to decrease the load on the heart, but congestive failure causes it to do just the opposite. Because the kidney correctly senses that it is getting less blood than normal, it compensates by producing hormones that actually cause reabsorption of the salt and water it has already filtered, so that they are returned to the circulation. The result is to increase the body's total fluid volume instead of decreasing it, thereby adding to the problems of the already-overworked heart. The failing heart thus outsmarts the kidney and itself at the same time; the self-same organ that is trying to be its friend becomes its inadvertent enemy.

Heavy, wet lungs with a sluggish circulation are an ideal breeding ground for bacteria and advancing inflammation, which is why so many cardiac patients die of pneumonia. But the heavy, wet lungs do not need the help of bacteria to do their killing. A sudden worsening of their waterlogged state, called acute pulmonary edema, is the frequent final event for patients with longstanding heart disease. Whether due to new cardiac damage or a temporary overload resulting from unexpected exercise or emotion, or perhaps just a little too much salt in a sandwich (I know of a man who died of what some might call acute pastrami-generated heart failure), the excessive fluid volume dams up and floods the lungs. Severe air hunger rapidly supervenes, the gurgling, wheezing respirations begin, and finally the poor oxygenation of the blood causes either brain death or ventricular fibrillation and other rhythm disturbances, from which there is no return. All over the world, at this very instant, there are people dying in this way.

The final passage of some of them is epitomized in the case history of another man whose death I witnessed. In the reference frame of chronic heart disease, Horace Giddens might be called Everyman. The details of his illness graphically depict one of the common patterns in the inexorable downhill course of cardiac ischemia.

Giddens was a successful forty-five-year-old banker in a small southern town when his path crossed mine in the late 1980s. He had just returned home from an extended stay at The Johns Hopkins Hospital in Baltimore, where his physician had sent him in desperation, hoping that the progression of his increasingly severe angina and heart failure might be slowed, or at least ameliorated; virtually every known treatment had already failed. Trapped in a strife-ridden marriage, Giddens had made the difficult journey to Baltimore as much to separate himself from the enervating enmity of his wife, Regina, as he had to seek some relief for his heart. But it was too late—his disease was found to be so far advanced that he was beyond help from any available therapy. After all the tests and consultations, the Hopkins doctors told him, as sensitively as they could, that even they could not help him—he was no candidate for any treatment other than palliative medication. For Horace Giddens, there would be no angioplasty, no bypass, no heart transplant. I was making a purely social visit to his home on the evening he returned from Baltimore courageously facing the certainty that he would soon be dead.

Although it was understood that Giddens was on his way home, his unfeeling wife seemed not to know or even care about the exact time of his arrival. When he actually entered the house, I was sitting quietly in a chair, listening to the family's conversation but not partaking in it. That entrance was a difficult moment to watch. The tall, gaunt Giddens came shuffling into the living room, grimacing with breathlessness, his narrow shoulders held firmly in the supporting grip of the adoring family maid. From a large photograph on the piano, I could tell that he had once been a robustly good-looking man, but now his grayish face was tired and drawn. He walked stiffly, as if with enormous effort, and carefully, seemingly unsure of his balance; he had to be helped into an armchair.

I knew of Giddens's history of angina, and I also knew that he had already sustained several full-blown myocardial infarctions. Watching the small shoulder-heaving struggle of each paroxysmal breath, I tried to imagine the condition of his heart and also attempted to put together in my mind's eye the various elements of the way it had failed him. After nearly forty years as a doctor, this kind of conjecture is a common preoccupation of mine when I find myself socially in the presence of the sick. It is an automatic drill, a self-testing, and in its own peculiar way, a kind of empathy as well. I do it always, almost without thinking. I'm sure many of my colleagues do the same.

What I visualized behind the breastbone of Horace Giddens was an enlarged, flabby heart that was no longer able to beat with anything resembling vigorous energy. More than three inches of its muscular wall had been replaced by a large whitish scar, and there were several other smaller areas of scarring as well. Every few beats, there was an irregular spasmodic contraction that originated from one or another rebellious focus on the left ventricle, intruding on the muscle's ineffectual attempt to maintain its steady rhythm. It was as though various parts of the ventricles were trying to break free of the intrinsic automaticity of the process, while the SA node struggled to maintain its declining authority. I knew the process well: The severity of the ischemia had cut off the regular messages that Giddens's SA node was trying to transmit to his ventricles. Unable to get their accustomed call, the ventricles feverishly begin to initiate beats on their own, starting each pulsation from whichever spontaneous spot on the myocardium chooses to meet the challenge. Any small increase in stress or decrease in oxygenation leads to a state of what the French so aptly call "ventricular anarchy," as disordered, ineffective contractions spread every which way through the heart muscle, giving way to the totally uncoordinated rapidity known as ventricular tachycardia and then fibrillation. As I watched Giddens's uncertain movements, I could easily tell how close he was to this series of terminal events.

The vena cavae and the pulmonary veins were distended and tense with the pressure of the blood backed up into them because of the heart's weakness. The leathery lungs resembled gray-blue water-soaked sponges, overloaded with puffy edema and barely able to rise and fall like the gentle pink bellows they once were. The whole blood-choked image reminded me of an autopsy I once saw of a man who had hanged himself—his livid purplish face was engorged and bulging, its plethoric features almost unrecognizable as human.

Giddens had lived his life well, and borne with philosophical resolve the slings and arrows fired at him by his malicious wife. He had devoted his life to the seventeen-year-old daughter who idolized him, and to the fulfillment of the trust put in him by the people of his town, whose admiration and respect he had earned by dint of simple probity and the wisdom of sound financial management of their savings. But now he had come home to die.

As I watched his nostrils flare with each difficult breath, I could not help but notice that the very tip of Giddens's nose was just a bit blue, and so were his lips—the wetness in his lungs was preventing proper oxygenation. The laboriously shuffling gait was the product of ankles and feet so swollen they seemed to bulge out over the tops of shoes made too small by the tightly constrained wet flesh within them. Every organ in the man's waterlogged body had some element of edema in it.

Pump failure was only part of the reason that walking was such an enormous effort for Giddens. He must have been agonizingly aware of the effort expended in each step he took, knowing that even the smallest increase in activity might bring on the dreaded pain of angina, since the hair-thin channels of his rigid coronary arteries were incapable of delivering any added requirement of blood.

Giddens sat down in the armchair and spoke briefly with his family, seemingly unaware of my presence. Tiring in both body and spirit, he then climbed laboriously up the staircase to his bedroom, stopping several times to look down and say a few words to his wife. As I watched him do this, I was reminded of a practice commonly resorted to by so-called cardiac cripples in order to disguise the advanced state of their illness: A patient feeling the onset of an anginal attack while on his daily stroll finds it useful to stop and gaze with feigned interest into a shop window until his pain disappears. The Berlin-born medical professor who first described the face- (and sometimes life-) saving procedure to me called it by its German name of *Schaufenster schauen*, or window-shopping. The *Schaufenster schauen* strategy was being used by Giddens to give him just enough respite to avoid serious trouble as he slowly made his way up to bed.

Horace Giddens died on a rainy afternoon only two weeks later. Although present, I was unable to lift a finger to help him. I could do nothing but sit by while his wife verbally abused him, until he suddenly threw his hand up to his throat, as though gesturing toward the brutal pathway of his radiating angina. His pallor suddenly increasing, he began to gasp, then shakily groped for the solution of nitroglycerin that lay on a coffee table in front of the wheelchair in which he sat. He managed only to get his fingers around it, but it fell from his trembling hand to the floor and shattered, spilling the precious medicine that might have widened his coronary arteries just enough to save him. Panic-stricken and breaking out into a cold sweat, he begged Regina to find the maid, who knew where his reserve bottle was kept. She didn't move.

Increasingly agitated, he tried to shout, but the only sound to come out of his mouth was a hoarse whisper, too small to be heard outside the room. The look on his face was heartrending to see, as he realized the futility of his strangled efforts.

I felt impelled to rush to Giddens's assistance, but something held me rooted to my chair. I didn't do a thing, and neither did anyone else. He made a sudden furious spring from his wheelchair to the stairs, taking the first few steps like a desperate runner trying with his last iota of energy to reach safety. On the fourth step, he slipped, gasped hungrily for air, seized the railing, and, in one great exhausted effort of grimacing finality, reached the landing on his knees. Frozen in my place, I gazed up the stairs at him and saw his legs give way. Everyone in that room heard the crumpling sound of his body falling forward, just out of view.

Giddens was still alive, but barely. Regina, with the calm dispatch of an experienced assassin, called out to two of the servants to carry him into his room. The family physician was summoned. Within a few minutes, and long before the doctor arrived, his stricken patient was dead.

Although I have assumed that the specific mechanism that killed Horace Giddens was ventricular fibrillation, it may have been acute pulmonary edema, or the terminal condition called cardiogenic shock, in which the left ventricle is just too weak to maintain a blood pressure high enough to sustain life. Among those of us who will succumb to ischemic heart disease, these three events will account for the vast majority of deaths. They can occur in sleep and they can happen so rapidly that only minutes pass before the moment of death. If medical help is at hand, the worst of their accompaniments can be lessened by morphine or other narcotics. The miracles of modern biomedicine can delay them for years. But every victory over ischemic heart disease is only a triumph of temporizing. The unremitting progression of atherosclerosis will continue, and each year there will be those more than half a million Americans who will die because the natural order demands it: Though it is a seeming paradox, natural death is the only way by which our species can be perpetuated.

By now, it may have become apparent why I was unable to lift a finger to help the unfortunate man who was dying before my eyes. I was watching the tragedy of Horace Giddens while comfortably seated in the seventh row of a theater, at a revival of Lillian Hellman's remarkable play *The Little Foxes*. Her clinically meticulous account of a fictional character dying of ischemic heart disease in 1900 could not have been more accurate had it been written by a cardiologist. Whole sentences of my description above are simply extracts from Miss Hellman's stage directions. The authoritative doctor who saw Giddens at Johns Hopkins was almost certainly the same William Osler whose words were quoted some pages ago.

Hellman's writing accurately portrays the very way in which so many of the victims of coronary ischemia still die today. For, in spite of all the delaying and comfort-enhancing tactics that modern medicine has produced in its battle against cardiac disease, the final scene in the struggle of a sick heart, now near the dawn of the twenty-first century, is often exactly like that in which Horace Giddens was the leading protagonist one hundred years ago.

Although many victims of ischemic heart disease still die in their first episode, like James McCarty, most follow a course more like that of Horace Giddens, in which the initial infarction or the evidences of ischemia are survived, then followed by a long period of careful living. In Giddens's time, careful living consisted of exactly what the term implies, a life free of physical or mental stress. Nitroglycerin was prescribed to abort angina, and a mild sedative given to allay anxiety. A certain therapeutic nihilism in vogue at the time among the university-based doctors may have been the reason they did not recommend the use of digitalis to increase the strength of the ventricle's contraction. Digitalis would not have prevented the coronary spasm that probably carried Giddens off, but it would certainly have lessened the chronic congestive failure from which he suffered so badly during his last months.

Nowadays, things are different. The spectrum of options available to treat ischemic heart disease mirrors the succession of accomplishments of modern biomedical science itelf, ranging from simple changes in lifestyle to the transplantation of a heart. Ischemia does its ruinous work in a variety of ways, and the myocardium needs help against every one of them. It is the job of the cardiologist to provide that help. In order to do that, he or she must know the nature of the enemy and the details of the strategy it is using in any given campaign. Specifically, the cardiologist begins by evaluating not only the current condition of the patient's heart and its coronary arteries but also the probability that worsening is so imminent that active steps must be taken to prevent it. To this end, a group of tests have been developed that are now so commonly utilized, their names and acronyms have become part of the common parlance of patients and their friends: Thallium stress test, MUGA, coronary angiogram, cardiac ultrasound, and Holter monitor are just a few.

Even with the objective information provided by such tests, it is impossible to give sound advice to a patient without understanding a great deal about his or her life and personality. It is not enough to measure the fraction of contained blood ejected by the ventricle with each contraction or to know the residual caliber of the narrowed coronary arteries, the mechanics of myocardial contraction, the output of the heart, the hypersensitivity to irritable stimuli of its electrical system, or any of those other factors so assiduously and impersonally determined in laboratories and X-ray units. The cardiologist must have a clear sense of the types of stresses existing in a patient's life and the likelihood that they can be changed.

Family history, dietary and smoking patterns, probability of compliance with medical advice, plans and hopes for the future, dependability of a support system of family and friends, personality type, and potential for modification if necessary—these are all factors that must be given proper weight in making decisions about treatment and long-term prognosis. It is the cardiologist's skill as a physician that enables him to befriend his patient and to know him—it is inherent in the art of medicine to appreciate that the testing and medications are of limited usefulness without the talking.

Having tested and talked, it is time to treat. Treatment is directed at decreasing the stress to which the heart is exposed, building up its reserve and resiliency for the long term, and correcting the specific abnormalities discovered during the process of testing. Implicit in all therapies is the necessity to do whatever is possible to slow the advance of atherosclerosis, recognizing that it can never be stopped entirely. Implicit also is the thesis that the heart is far more than just another stolidly stupid pump—it is a responsive, dynamic participant in the enterprise of life, capable of adaptation, accommodation, and, to some extent, repair.

William Heberden, without knowing it, described in 1772 what we may now recognize as a classic example of the way in which a properly designed exercise program may build up the heart's ability to respond to those challenging moments when increased work is demanded of it. Writing of patients with angina, he reported: "I know one who set himself a task of sawing wood for half an hour every day, and was nearly cured." Although the handsaw has nowadays been replaced by the stationary bicycle, the principle is the same.

A wide variety of cardiac medications is available to help the heart muscle and its conduction system in their resistance to the effects of ischemia, and most assuredly there will be more. There are even drugs that may be used within the first few hours of a coronary occlusion, to dissolve the brand-new clot that has caused the final bit of obstruction in the atherosclerotic vessel. There are drugs to decrease myocardial irritability, prevent spasm, dilate coronary arteries, strengthen the heartbeat, diminish accelerations of rate, drive out the excess load of water and salt in congestive failure, slow down the clotting process, decrease cholesterol levels in the blood, lower blood pressure, allay anxiety—and every one of them carries with it the possibility of undesirable or frankly dangerous side effects, for whose treatment, of course, there are still other drugs. Cardiologists of today tread a fine line between so drying out a patient that he is too weakened to live normally, and allowing him so much of a fluid load that he is in danger of lapsing into serious congestive failure.

In no area of human infirmity have the wizardries of electronics contributed so much as in the management of heart disease. Although diagnosis has been the primary beneficiary of their miracles, therapy, too, has been enhanced by the physicists and engineers who deal in such esoterica. We now have pacemakers to do the job of the SA node; they safely trigger a predictable and steady beat. There are defibrillators that not only reassert control when the heart's mechanism becomes frivolous but even have the added virtue of being implantable directly into the patient, so that response to irregular rhythm is automatic and instantaneous.

Surgeons and cardiologists have devised operations to reroute blood around obstructions in coronary arteries and to widen narrowed vessels with balloons, techniques known respectively as the coronary artery bypass graft, or CABG (predictably pronounced *cabbage*), and angioplasty. When all else fails, an occasional patient fulfills the criteria to have his whole heart chucked out and replaced with a healthy secondhand one. All these operations, when the candidate is carefully chosen, have high rates of success. And yet, after each one, the process of atherosclerosis continues to lick at life. Widened arteries frequently plug up again, grafted vessels develop atheromata, and the symptoms of ischemia too often return to their old myocardial haunts.

Delay it though we may, then, the victims of coronary atherosclerosis will almost certainly die of their affliction—perhaps unexpectedly during a time when they seem to be responding well to treatment, perhaps of the gradual effects of congestive heart failure. Although its more flagrant symptoms are less commonly seen than they were in the days before the advent of effective ways to fend them off, chronic congestive failure remains a significant force in the demise of many people with ischemic heart disease. Once the heart has become so weakened that congestive failure occurs, the outlook is poor. Approximately half its victims will die within five years. As we noted earlier, along with the sharp drop in actual heart attacks in recent years has come a dramatic rise in the incidence of failure, a rise that will probably continue. There are now many more Horace Giddenses and many fewer James McCartys.

The reasons for this are several. The most obvious is that not only physicians but also community facilities have considerably improved their ability to cope with the urgent situations created by myocardial infarction. Speedy response by highly skilled paramedics and efficient transfer to an emergency room have meant better treatment during the first crucial hours, and the in-hospital intensive care itself has vastly improved. But another factor is at least equally important. More effective methods of medical care in general have resulted in the survival of increasing numbers of people to an older age, an age at which weakened cardiac pumping and consequent congestive failure are more commonly a problem.

The incidence of heart failure in people younger than fifty-five has actually dropped—the great increase in overall numbers occurs entirely in the population older than sixty-five. More than 2 million Americans have some degree of heart failure that restricts their activities and undermines their vitality. When it becomes severe, it carries a mortality rate of 50 percent in two years. Thirty-five thousand people will die of it annually, far fewer than the 515,000 who will succumb to an actual heart attack, but a large number nevertheless.

Those whose hearts do not quit because of ventricular fibrillation and arrest will eventually die for the reasons previously enumerated: They cannot breathe well enough to oxygenate blood, their kidneys or liver can no longer cleanse their bodies of toxic substances, bacteria run rampant through their systems, or they simply cannot sustain a blood pressure high enough to maintain life and, most particularly, the function of the brain. This last-named condition is called cardiogenic shock. It and pulmonary edema are by far the most common cardiac enemies that are perpetually being fought in intensive care units and emergency rooms. The patients and their medical allies will win most of those battles, at least temporarily.

Having countless times watched those teams fighting their furious skirmishes, and having often been a participant or their leader in years past, I can testify to the paradoxical partnering of human grief and grim clinical determination to win that actuates the urgencies swarming through the mind of every impassioned combatant. The tumultuous commotion of the whole reflects more than the sum of its parts, and yet the frenzied work gets done and sometimes even succeeds.

As chaotic as they may appear, all resuscitations follow the same basic pattern. The patient, almost invariably unconscious because of inadequate blood flow to the brain, is quickly surrounded by a team whose mission is to pull him back from the edge by stopping his fibrillation or reversing his pulmonary edema, or both. A breathing tube is rapidly thrust through his mouth and down into his windpipe so that oxygen under pressure can be forced in to expand his rapidly flooding lungs. If he is in fibrillation, large metal paddles are placed on his chest and a blast of 200 joules is fired through his heart in an attempt to stop the impotent squirming, with the expectation that a regular beat will return, as it frequently does.

If no effective beat appears, a member of the team begins a rhythmic compression of the heart by forcing the heel of his hand down into the lowest part of the breastbone at a rate of about one stroke per second. By squeezing the ventricles between the flatness of the yielding breastbone in front and the spinal column in back, blood is forced out into the circulatory system to keep the brain and other vital organs alive. When this form of external cardiac massage is effective, a pulse can be felt as far away as the neck and groin. Although one might think otherwise, massage through an intact chest results in far better outcomes than does the direct manual compression that was the only known method when I had my grim encounter with the obstinacy of James McCarty's myocardium some forty years ago.

By this point, IVs will have been inserted for the infusion of cardiac drugs, and wider plastic tubes called central lines are being expeditiously inserted into major veins. The various drugs injected into the IV tubing have assorted purposes: They help to control rhythm, decrease the irritability of the myocardium, strengthen the force of its contraction, and drive excess fluid out of the lungs, to be excreted by the kidney. Every resuscitation is different. Though the general pattern is similar, every sequence, every response to massage and drugs, every heart's willingness to come back—all are different. The only certainty, whether spoken or not, is that the doctors, nurses, and technicians are fighting not only death but their own uncertainties as well. In most resuscitations, those uncertainties can be narrowed down to two main questions: Are we doing the right things? and, Should we be doing anything at all?

Far too often, nothing helps. Even when the correct answer to both questions is an emphatic *yes*, the fibrillation may be beyond correction, the myocardium unresponsive to the drugs, the increasingly flabby heart resistant to massage, and then the bottom falls out of the rescue attempt. When the brain has been starved of oxygen for longer than the critical two to four minutes, its injury becomes irreversible.

Actually, few people survive cardiac arrest, and even fewer among those already seriously ill people who experience it in the hospital itself. Only about 15 percent of hospitalized patients below the age of seventy and almost none of those who are older can be expected to be discharged alive, even if the CPR team somehow manages to succeed in its furious efforts. When an arrest occurs elsewhere than the hospital, only 20 to 30 percent survive, and these are almost always those who respond quickly to the CPR. If there has been no response by the time of arrival in the emergency room, the likelihood of survival is virtually zero. The great majority of the responders are, like Irv Lipsiner, victims of ventricular fibrillation.

The tenacious young men and women see their patient's pupils become unresponsive to light and then widen until they are large fixed circles of impenetrable blackness. Reluctantly the team stops its efforts, and the entire scene becomes transformed from a vital image of imminent heroic rescue to the dejected gloom of failure.

The patient dies alone among strangers: well-meaning, empathetic, determinedly committed to sustaining his life—but strangers nonetheless. There is no dignity here. By the time these medical Samaritans have ceased their strenuous struggles, the room is strewn with the debris of the lost campaign, more so even than was McCarty's on that long-ago evening of his death. In the center of the devastation lies a corpse, and it has lost all interest for those who, moments earlier, were straining to be the deliverers of the man whose spirit occupied it.

What has happened is the culmination of a straightforward series of biological events. Whether programmed by his genes, self-imposed by the habits of his life, or, as is usually the case, a combination of both, a man's coronary arteries have been unable to bring sufficient blood to nourish the muscle of his heart; the heartbeat became ineffective, the brain went too long without oxygen, and the man died. Approximately 350,000 Americans suffer a cardiac arrest each year, and the vast majority of them die; fewer than one-third of the episodes occur in a hospital. Often, there is no warning of the imminence of that final exit. No matter how much ischemia a heart has endured in the past, its defection may be sudden. In some 20 percent of people, it may even happen, as it did for Lipsiner, without pain. Whatever mystery attaches to such a death is imposed on it by those who live. It is a tribute to the human spirit that the life preceding triumphs over the ugly events that most of us will experience as we die, or as we move toward our last moments.

The experience of dying does not belong to the heart alone. It is a process in which every tissue of the body partakes, each by its own means and at its own pace. The operative word here is *process*, not *act*, *moment*, or any other term connoting a flyspeck of time when the spirit departs. In previous generations, the end of the faltering heartbeat was taken to indicate the end of life, as though the abrupt silence beyond it intoned a soundless signal of finality. It was a specified instant, recordable in the chronicle of life and marking a full stop after its concluding word.

Today the law defines death, with appropriate blurriness, as the cessation of brain function. Though the heart may still throb and the unknowing bone marrow create new cells, no man's history can outlive his brain. The brain dies gradually, just as Irv Lipsiner experienced it. Gradually, too, every other cell in the body dies, including those newly alive in the marrow. The sequence of events by which tissues and organs gradually yield up their vital forces in the hours before and after the officially pronounced death are the true biological mechanisms of dying. They will be discussed in a later chapter, but first it is necessary to describe the prolonged form of dying that is old age.

III

Three Score and Ten

NO ONE DIES of old age, or so it would be legislated if actuaries ruled the world. Every January, just when the harsh autocracy of winter has tightened its hoary hold, the U.S. government releases its yearly "Advance Report of Final Mortality Statistics." Neither among the top fifteen causes of death nor anywhere else in that soulless summary is there to be found a listing for those among us who just fade away. In its obsessive tidiness, the Report assigns the specific clinical category of some fatal pathology to every octo- and nonagenarian in its neat columns. Neither do those few whose age is recorded in three digits escape the orderly nomenclature of the tabulators. Everybody is required to die of a named entity, by order not only of the Department of Health and Human Services but also of the global fiat of the World Health Organization. In thirty-five years as a licensed physician, I have never had the temerity to write "Old Age" on a death certificate, knowing that the form would be returned to me with a terse note from some official record-keeper informing me that I had broken the law. Everywhere in the world, it is illegal to die of old age.

Actuaries seem unable to accept a natural phenomenon unless it is so well defined as to fit neatly into a distinct and easily described classification. The annual report of the federal death accountants is very orderly—not very imaginative, and to my mind not completely reflective of real life (and real death), but nevertheless very orderly. I'm convinced that plenty of people do die of old age. Whatever scientific diagnoses I have been scribbling on my state's death certificates to satisfy the Bureau of Vital Statistics, I know better.

At any given moment, some 5 percent of our nation's elderly reside in long-term care facilities. If they have been there longer than approximately six months, the vast majority of them will never leave these nursing homes alive, except perhaps for a brief terminal sojourn in a hospital, where some young house physician will eventually fill out one of those very proper death certificates. What do all of these old people die of? Though their doctors dutifully record such distinct entities as stroke, or cardiac failure, or pneumonia, these aged folk have in fact died because something in them has worn out. Long before the days of scientific medicine, everyone understood this. On July 5, 1814, when he was seventy-one years old, Thomas Jefferson wrote to the seventy-eight-year-old John Adams, "But our machines have now been running seventy or eighty years, and we must expect that, worn as they are, here a pivot, there a wheel, now a pinion, next a spring, will be giving way; and however we may tinker them up for a while, all will at length surcease motion."

Whether its overt physical manifestation appears in the cerebrum or in the sluggishness of a senile immune system, the thing that peters out is nothing other than the life force. I have no real quarrel with those who insist upon invoking the laboratory-bred specificity of microscopic pathology in order to satisfy the compulsive demands of their biomedical worldview—I simply think they miss the point.

As soon as I became conscious of life, I began the long process of watching someone gradually die of old age. The statistician has not yet drawn breath who can convince me that my grandmother's state-certificated Cause of Death was anything else than a legalized evasion of the greater law of nature. She was 78 years old at my birth, although her yellowed immigration papers claimed she was only 73—twenty-five years earlier at Ellis Island she had chosen to be younger than truth would allow, because she had been told that the figure 49 would be more acceptable than 54 to the stern, soldierly AMERICAN official in the brass-buttoned uniform who asked those blunt questions she assumed were so crucial to her entry. So you see, I am not the first member of my clan whose fear of governmental rejection has led to a bit of perjury.

Three generations of my family shared a four-room apartment in the Bronx, six souls altogether—my grandmother, my maiden aunt Rose, my parents, my older brother, and I. In those days, it was unthinkable that an aged parent would be sent off to one of the few available Homes. Even if the will existed, which it rarely did, there was simply no way. Half a century ago, for people like us to so banish an aged parent was considered a coldhearted circumvention of responsibility, and a denial of love.

My high school was only half a block from our tenement house, and even the distance to my college was a walk of no more than twenty minutes. Each morning, my grandmother put a small sandwich and an apple into a brown paper bag, which I then squashed between my books and forearm as I went off to the verdant campus on the hill. At intervals along the way, I would be joined by chums whom I had known since PS 33. By the beginning of my second class of the morning, the bag had invariably become greasy from the thickly layered butter my adoring grandmother always slabbed too generously onto the Silvercup. To this day, I can't see an oleaginous smear on brown paper without feeling the sweet pain of nostalgia rise up in my chest.

Very early every day, my aunt Rose and my father disappeared into the subway that took them to their dressmaking jobs in the garment section of Manhattan. My mother died when I was eleven, and I was my grandmother's boy. Except for an appendicitis hospitalization, and two half-month periods when a monied relative financed brief intervals at a summer camp, I spent most of every day of my life in her close company. Without realizing it, I passed my first eighteen years watching her descent to death.

When six people live in an apartment of four small rooms, there are very few secrets. During her last eight years, my grandmother shared a bedroom with my aunt and me. Till the day I completed my last college assignment, my homework was done on a card table opened up in the center of our small living room, while the activities of the household went on within a few feet of me. When my studying was over, I would fold the table and its collapsible chair and stow them against the wall behind the open door leading from the short entrance foyer into the living room. If I left so much as a scrap of paper behind, I would hear about it from my grandmother.

"Grandmother" was not the name we used for the matriarch of our family, because "Grandmother" spoke only a few monosyllabic words of English. What my brother and I called her was the Yiddish equivalent, "Bubbeh," and what she called us was Herschel (my brother's name was Harvey) and Shepsel. To this day, everyone calls me Shep, and it is a memory of my Bubbeh.

Bubbeh's life had never been easy. Like so many Eastern European immigrants, her husband had preceded her to the golden shores of America and had taken their two sons with him, leaving his wife for several years with four young daughters in a tiny Belarusian village. Only a few years after the family's life was reconstituted in a crowded (because other relatives shared it) apartment on Rivington Street on New York's Lower East Side, my grandfather and both boys died in quick succession, whether from tuberculosis or influenza is uncertain.

By this time, three of the four daughters were working in garment-industry sweatshops, so there was some money coming in. Taking advantage of a subsidy made available by a Jewish philanthropy, Bubbeh scraped together enough dollars to make a down payment on a two-hundred-acre farm near Colchester, Connecticut, and joined a large group of her countrymen who were doing the same. Like the others, she worked the land with the help of a series of hired men, one after the other, usually Polish immigrants who spoke no more English than she did. How this four-foot-ten-inch iron-willed dynamo survived that period is difficult to know, because the farm was not very productive. Its real income, barely adequate for day-to-day expenses, came from the small contributions of family and old-country friends who stayed at the place for short periods to escape the tuberculosis-threatened closeness of lower Manhattan's Tenth Ward.

As their source of refuge and a wellspring of fortitude against the confusing hurly-burly of America, Bubbeh took on the role of what I can only describe as a Yiddishe *mater et magistra* to an extended group of struggling young immigrants. Though she could not speak a comprehensible sentence of English, she somehow understood the rules and the rhythms of American life. If there were "wonder rabbis" in the old country, the enlarging clan had found a female fount of oracular stature in this new one, and bestowed upon her the honorific *tante*, or "aunt." As Tante Peshe, which translates only inadequately into Aunt Pauline, her strength embraced a large and needy congregation of self-appointed nephews and nieces, some of whom were barely younger than she.

Eventually, the farm had to be given up when all but one of the girls married. But long before then, the eldest daughter, Anna, had died in her twenties of childbed fever, and her young husband had gone off to pursue his own life. Left in mourning with Anna's baby, Bubbeh raised him on the farm as her own son. He was in his late teens when the farm was sold, and the Bronx period of our family life began.

By the time I was eleven, my aunt Rose was my grandmother's only surviving child. One had died during infancy, and the others in this land to which they had brought their dreams. Bubbeh was then eighty-nine years old, a tiny, exhausted figure, who kept life's fires aglow, barely, for the sake of her three young grandchildren: my brother and me, as well as my thirteen-year-old cousin, Arline. Arline had come to us two years earlier when her mother died of kidney failure; she later left to live with her father's family when my mother succumbed to cancer shortly after my eleventh birthday. The record of Bubbeh's long widowhood was an unrelieved chronicle of struggle, sickness, and death. Her hopes had been lowered one by one into the grave with her husband and six children. There remained only my aunt Rose and the three of us born in the land whose promise had turned to heartbreak.

It must have been after my mother died that I first began to be conscious of just how ancient Bubbeh was. Since earliest memory, I had amused myself from time to time by playing idly with the loose, unresilient skin on the backs of her hands or near her elbows, gently drawing it out like stretched taffy, then watching in never-lessening wonder as it slowly resettled into place with an easy languor that made me think of molasses. She would slap my hand sharply when I did this, in mock annoyance at my boldness, and I would laugh teasingly until her eyes betrayed her own amusement at my feigned disrespect. In truth, she loved my touch, as I loved hers. Later, I became aware that I could produce a shallow pit in the tissues of her shin simply by pressing the lisle-stockinged skin hard against the bone with my fingertip. It took a long time for the pit to fill back in and disappear. Together, we would sit silently and watch it happen. With time, the pits became deeper and the filling-in period grew longer.

Bubbeh moved from room to room in slippered feet and with great care. As the years passed, the walk became a shuffle, and finally a kind of slow sliding, the foot never leaving the floor. If for any reason she had to move a bit faster, or if she was upset at one of us kids, she became short-winded, seeming to find it easier to breathe if she opened her mouth widely to draw in the air. Sometimes she let her tongue hang forward just a bit over her lower lip, as though in hope of absorbing some extra oxygen through its surface. I didn't know it, of course, but she was beginning the gradual slide into congestive heart failure. Almost certainly, the failure was aggravated by the significant decline in the amount of oxygen that aged blood is capable of taking up from the aged tissues of the aged lung.

Slowly, her vision, too, began to fail. At first, it became my job to thread her sewing needles, but when she found herself unable to guide her fingers, she stopped mending altogether, and the holes in my socks and shirts had to await the few free evening moments of my chronically fatigued aunt Rose, who laughed at my puny attempts to teach myself to sew. (In retrospect, it seems hardly possible that I would one day be a surgeon; Bubbeh would have been very proud, and very surprised.) After some years, Bubbeh could no longer see well enough to wash dishes or even sweep the floor, because she couldn't tell where the dust and dirt were. Nevertheless, she wouldn't give up trying, in a futile effort to retain even this small evidence of her usefulness. Her persistent attempts to clean became a source of some of the small daily frictions that must have made her feel increasingly isolated from the rest of us.

In my early teens, I saw the last traces of the old combativeness disappear and my grandmother became almost meek. She had always been gentle with us kids, but meekness was something new—perhaps it was not so much meekness as a form of withdrawal, an acquiesence to the expanding power of the physical disablements that were subtly increasing her separation from us and from life.

Other things began to happen. In time, Bubbeh's decreasing mobility and unsteadiness made it impossible for her to get to the bathroom at night, and so she slept with a large Maxwell House coffee can under the bed. Most nights, I would be awakened by her awkward attempts to find it in the dark or by the sound of her weak stream hitting its tinny inside. Many were the times I lay motionless in the predawn blackness, peering across the room at Bubbeh crouched uncomfortably alongside her bed, the coffee can held high under her nightgown with one unsteady hand as she tried to stabilize her tottering body against the mattress with the other.

I could never understand why Bubbeh had to get up so often for those nocturnal bouts with the coffee can, until many years later when I learned of the marked reduction in bladder capacity that occurs with age. Unlike many old people, Bubbeh never became incontinent, although I'm sure there were minor episodes I never knew about. Not until her very last months was she sometimes betrayed by the faint odor of urine, but even then only when I stood very close or hugged her frail form snugly against my body.

Bubbeh lost her last teeth when I was in early adolescence. She had saved all of them in a small change purse kept toward the back of the top drawer of a bureau she and Aunt Rose shared. One of the secret rituals of my childhood was to sneak into the drawer and gaze for a few moments in awe at those thirty-two yellow-white objects, no two of which seemed alike. To me, they were so many little milestones of my grandmother's aging and the history of our family.

Even without teeth, Bubbeh somehow managed to eat most food. Toward the end, she lacked the strength even for that, and her nutrition suffered. The inadequate intake added to the usual decrease that aging causes in muscle mass, and it changed the configuration of her body, making her seem shriveled in comparison with the stalwart, slightly stoutish old lady I had once known. Her wrinkling increased, her complexion faded into a uniformity of mild pallor, the skin of her face seemed to hang ever more loosely, and the old-world beauty she had retained until her nineties was finally lost.

There are simple clinical explanations for the many things I saw during my grandmother's declining years, but somehow they seem unsatisfactory even now. It is all very well to speak of such causative factors as decreased circulation to the brain or senescent degeneration in cerebral cells so subtle that electron microscopy is needed to demonstrate it—there is a certain intellectual detachment to be found in the stark biological description of the death of that very tissue that once enabled a nonagenarian to think clear and sometimes audacious thoughts. The researches of physiologists might here be cited, as well as the work of endocrinologists, psychoneuroimmunologists, and the rapidly evolving modern breed of gerontologists, to explain everything that was unfolding before my adolescent eyes. But that actual seeing is what demands attention, the seeing of a process in the midst of which all of us constantly live. Surrounded by it though we may be, there is for each of us that something within that turns the face of consciousness away from the reality of our own concomitant aging. Something within us will not accept the immediacy of awareness that, even as we bear witness to it in the obviously old, our own bodies are simultaneously and subtly undergoing the same inexorable process that will lead eventually to senescence and death.

And so my grandmother's brain cells had long before this time begun to die, even as mine are dying today, and yours. But because she was much older than I am just now, and because she was withdrawing from the stimulation of the world around her, her decreased number of brain cells and their decreased responsiveness led to very obvious changes in her behavior. Like all old people, she became increasingly forgetful and was annoyed when reminded of it. Always known for the forthrightness of her dealings with people, she grew overtly irritable and impatient with those few outside our immediate family with whom she still came into contact, and she seemed to rouse herself by offending even those who had in years past looked to her for guidance. Then the time came when she began to sit silently even in company. Eventually, she spoke only when she absolutely had to, distantly and with little emotion.

What was most evident, but only in retrospect, I confess, was the slow drawing away from life. When I was a small boy, and even into my early teens, my grandmother prayed in the synagogue on the High Holy Days. As difficult as the five-block pilgrimage became, she would somehow manage it, negotiating cracked areas in the Bronx pavement with her worn prayer book held very snugly under one armpit lest she sin by letting it fall to the ground. I would take her there. How I regret every murmur of complaint, how I wish I was not sometimes—no, not sometimes, but often—ashamed of being seen with this black-kerchiefed, shuffling remnant of a shtetl culture that was all but gone even as she stubbornly refused to join it in the grave. Everyone else's grandparents seemed so much younger, they spoke English, and they were independent—mine was a reminder not only of the lost world of Eastern European Jewry but of my own turbulent conflict about the load of emotional detritus I nowadays euphemistically call my heritage.

With her free hand, Bubbeh would hold tightly onto my arm, sometimes gripping the cloth of my sleeve as I guided her with agonizing slowness through the streets, then down the stairs into the synagogue's vestry (our family prayed in the cheap seats, and could barely afford even those), and then to her chair amidst other ladies we called elderly, but very few of them nearly as foreign or as careworn as she. A few moments later, I would leave her there, her head already bowed over the old tearstained book she had prayed from since girlhood. Its words were printed in both Hebrew and Yiddish, but she prayed from the Yiddish side of the page because it was the only language she knew. Through the long ritual of those holiday services, she quietly murmured the words that became with each passing year more laborious to read, and finally impossible. About five years before her death, Bubbeh could no longer make the long walk to the synagogue, even with both grandsons to help her. Relying largely on her still-intact long-term memory, she recited the liturgy at home, sitting by the open window as she had done every Saturday morning during all the years I knew her. After a few years, even that became too much. She could barely see the sentences and her memory for the prayers learned in her youth was giving out. Finally, she stopped praying altogether.

By the time Bubbeh stopped praying, she had stopped virtually everything else as well. Her food intake had become minimal—she spent most of each day seated quietly at her window, and she spoke sometimes of death. And yet she had no disease. I'm sure some eager physician might have pointed out her chronic cardiac failure and added to it the probability that there was an element of atherosclerosis, and perhaps he would have prescribed some digitalis. To me, that would have been like dignifying the degeneration of her joints by calling it osteoarthritis. Of course it was arthritis, and of course she was in chronic failure, but only because her pinions and springs were giving way under the weight of the years. She had never been sick a day in her life.

The government statisticians and the scientific clinicians insist that proper names must be applied to sluggish circulation and an antique heart. I have no quarrel with that, so long as they do not also insist that assigning a name to a natural biological state means *a priori* that it is a disease. Like the nerve cell, the muscle cell of the heart is one of those that cannot reproduce—as it gets older, it simply wears out and dies. The biological processes that throughout life have been making replacement parts for dying structures within each cell can no longer do their job. The mechanism by which a newly generated piece of cell membrane or intracellular structure can take the place of a section dead of too much use finally becomes inoperative. After a lifetime of regenerating spare parts, the nerve and muscle cells' capacity of rejuvenation gradually shuts down. The tactic of continuous renewal within each heart-muscle cell is then defeated by the overwhelming strategy through which aging is achieving its ultimate objective of destruction. One after another, like my grandmother's teeth, the cardiac muscle cells cease to live—the heart loses strength. The same process takes place in the brain and the rest of the central nervous system. Even the immune system is not immune to aging.

Changes that are at first only biochemical and intracellular become manifest in the function of entire organs. There is a gradual decrease in the cardiac output while at rest, and when the heart is stressed by exercise or emotion, its ability to increase is less than required by the needs of arms, lungs, and every other structure of the body. The maximal rate attainable by a perfectly healthy heart falls by one beat every year, a figure so reliable that it can be determined by subtracting age from 220. If you are fifty years old, it is unlikely that your heart can manage much more than 170 beats per minute, even under the most extreme conditions of emotion or exercise. These are only some of the ways in which the aging and stiffening myocardium loses its ability to adapt to the challenges presented to it by everyday life.

The rapidity of circulation slows down. The left ventricle takes longer to fill and longer to relax after a contraction; each heartbeat pushes out less blood than it did a year earlier, and even a smaller fraction of its content. Perhaps in an attempt to compensate, the blood pressure tends to rise somewhat. Between the ages of sixty and eighty, it increases by 20 millimeters of mercury. One-third of people over the age of sixty-five have hypertension.

Not only the muscle of the heart but also its conduction system dies out as the decades pass. By the age of seventy-five, the sinoatrial node may have lost as much as 90 percent of its cells; the bundle of His contains fewer than half of its original fibers. There are related electrocardiographic changes that go along with all of this loss of muscle and nerve tissue, and they can easily be identified on the inked tracing.

As the pump ages, its inner lining and valves thicken. Calcifications appear in the valves and muscle; the color of the myocardium changes somewhat as a yellow-brown pigment called lipofuscin is deposited in its tissues. Like the face of a weather-beaten old man, a heart looks its age. It acts its age as well. There is no need to invoke a disease to explain its failure. Cardiac failure is ten times as common in people older than seventy-five than in those between forty-five and sixty-five years of age. That is precisely the reason I could so easily indent the tissues of my grandmother's skin, and it was unquestionably the source of her easy shortness of breath. It is probably also the reason that the most common symptom of a heart attack in elderly patients is severe failure, rather than the classic picture of unremitting chest pain.

Not only the heart itself but the blood vessels, too, are affected by the passing years. The walls of the arteries thicken. Like the person in whom they dwell, they lose their elasticity; they can no longer constrict and dilate with the enthusiasm of youth, and so it is more difficult for the body's regulatory mechanisms to control the amount of blood going to muscles and organs to satisfy their ever-changing needs. Moreover, atherosclerosis continues on its inexorable way with each passing year. Even without the cholesterol-rich obesity or the cigarettes or diabetes that make it appear at a younger stage, the arterial walls gradually narrow as more and more atheroma accumulates with the prolonged contact of decade upon decade of coursing blood.

Before long, every organ is getting less nourishment than it needs to do the job intended for it by nature. Total blood flow to the kidney, for example, decreases by 10 percent for every decade after the age of forty. Actually, only some of that organ's decline is caused by the lessened cardiac output and narrowed vessels, but those factors worsen the effect of certain aging changes within the kidney itself. For example, between the ages of forty and eighty, the normal kidney loses some 20 percent of its weight and develops areas of scarring within its substance. Thickening of the tiny blood vessels that are inside the kidney further decreases blood flow and results in destruction of the organ's filtering units, which are, of course, the crux of its ability to clear the urine of impurities. In time, some 50 percent of the filtering units will die.

The changes in its structure decrease the kidney's effectiveness. With increasing age, it loses its ability not only to get rid of excess sodium but even to retain it in the body when needed. The result is an instability of the aged person's salt and water volume, tending to increase the possibility of either heart failure on the one hand or dehydration on the other. This is one of the main reasons that cardiologists treating the elderly have so much difficulty treading the narrow frontier between the Scylla of sodium overload and failure and the Charybdis of parched old tissues.

The result of all of these inadequacies is an increased propensity of the kidney to default in its responsibilities. Even when it does not fail outright but merely falters, it recovers more slowly than a younger organ, and is more prone to let its host down altogether under severe stress—death from kidney failure is a common pathway of exit when an aged person is weakened by some other pathology, such as late-stage cancer or liver disease. The blood's impurities build up; the other organs, particularly the brain, are poisoned; and death from so-called uremia is inevitable, often preceded by a variable period of coma. The terminal event in uremic patients is most commonly an irregularity of cardiac rhythm (an arrhythmia) caused by the kidney's inability to rid the blood of excess potassium. Victims of kidney failure usually slide into it imperceptibly, then are suddenly dead in a flash of cardiac instability. Only rarely are there any last words or deathbed reconciliations.

Although the kidney is the most significant part of the urinary tract to develop changes with age, the bladder, too, is affected. The bladder is essentially a thick balloon whose wall is made of flexible muscle. As it ages, the balloon loses its distensibility and can no longer hold as much urine as before. Old people need to urinate more frequently, and this is the reason my grandmother got up once or twice each night to grapple in the dark with her coffee can.

Aging also affects the finely tuned coordination between the bladder muscle and the shutter mechanism whose function is to keep urine from leaking out. The result is the occasional incontinence of the aged, which in some people becomes a major problem, especially if complicated by infection, prostate trouble, mental confusion, or medication. Disturbances of the bladder's emptying capacity are often a major factor in producing urinary-tract infections, a dangerous enemy of the debilitated elderly.

Like the muscle of the heart, brain cells are unable to reproduce. They survive decade after decade because their various structural components are always being replaced as they wear out, like so many ultramicroscopic carburetors and plugs. Though cell biologists use more abstruse terminology than do mechanics (words like *organelle* and *enzyme* and *mitochondrium*), these entities nonetheless require just as efficient a replacement mechanism as do their more familiar automotive analogues. Like the body itself and like each of its organs, every cell has the equivalents of pinions and wheels and springs. When the mechanism to exchange the aging parts for new wears out, the nerve or muscle cell can no longer survive the constant destruction of components that goes on within it.

That parts-replacement mechanism requires the participation of certain molecular structures within the cell. But the molecules in biological systems have a finite life span. Beyond that prescribed period, their constant collisions against one another change their character enough so that they can no longer generate new spare parts. By the process of wear and tear, they reach the limits of their longevity, thus limiting the longevity of the brain cell they serve. This is the biochemical process that scientists call cellular aging. The cell gradually dies and its fellows do the same. When enough of them are gone, the brain begins to show its age.

For every decade after age fifty, the brain loses 2 percent of its weight. When my Bubbeh died at ninety-seven, her brain weighed some 10 percent less than it did when she arrived in this country. The gyri, those twisting, raised convolutions in the cortex within which we do so much of the receiving and thinking that makes us different from the rest of God's creatures, suffer the greatest atrophy and loss of prominence. Concomitantly, the valleys between them (the sulci) become larger, as do the fluid-filled chambers deep in the brain's substance, known, like those in the heart, as ventricles. As though it were a biological marker of advancing senescence, lipofuscin stains the cells of white and gray matter alike, imparting to the shrinking brain a creamy yellow tinge that deepens with advancing age. Even senescence is color-coded.

As obvious as are the grossly visible changes in the withering brain, it is in microscopic appearance that aging is most evident. Particularly striking is the decrease in the number of nerve cells, or neurons, that results from the lethal spare-parts failure just described. The events that take place in the cortex are representative of the whole. The motor area of the frontal cortex loses between 20 and 50 percent of its neurons; the visual area in the back loses about 50 percent; the physical sensory part on the sides also loses about 50 percent. Fortunately, the higher intellectual areas of the cerebral cortex have a significantly lower degree of cell disappearance, much of which seems to be compensated for by overlap and redundancy of function. It may even be that the fewer neurons increase their activity, but whatever the reason, such intellectual capacities as reasoning and judgment are quite often unimpaired until late senescence.

Interestingly, recent research suggests that certain cortical neurons seem actually to become more abundant after maturity has been reached, and these cells reside in precisely the areas in which the processes of higher thought take place. When these findings are added to the confirmed observation that the filamentous branchings (called dendrites) of many neurons continue to grow in healthy old people who don't have Alzheimer's disease, the possibilities become quite intriguing: Neuroscientists may actually have discovered the source of the wisdom which we like to think we can accumulate with advancing age.

Except in highly localized areas, therefore, the cortex not only loses neurons but almost all of those it does retain exhibit signs of aging, as replacement of intracellular parts becomes gradually less efficient. The end result is that the brain is smaller than it was in youth, and doesn't work as well. In everyday life, this is manifest in all of that multitude of slowings that we see daily in our elderly associates, and too soon in ourselves. The brain is thus sluggish in its function and sluggish as well in its ability to spring back from biological insult—it recovers less efficiently from events that threaten its survival.

One of the most dangerous of those events is an interference with blood supply. When blood flow to some specific region of the brain is cut off (a catastrophe that usually happens suddenly), there is immediate dysfunction or death of the nerve tissue supplied by the obstructed artery. This is precisely what is meant by the term *stroke*. Strokes may occur for any of a number of reasons, but the most common among the elderly is atherosclerosis blocking branches of the two large vessels that nourish the brain, the right and left internal carotid arteries. Approximately 20 percent of hospitalized stroke victims die soon after the episode and another 30 percent require long-term or institutional care until death.

Though the death certificates of stroke victims have often been adorned with such terms as *cerebrovascular accident* or *cerebral thrombosis* (these days, the proper word is the simpler and all-encompassing *stroke*), more significant than the nomenclature on the legal paper is the number written into the blank space for *Age*: It is almost always high. Men and women beyond the age of seventy-five suffer ten times the incidence of strokes as do those between fifty-five and fifty-nine.

"Cerebrovascular accident," in fact, was what was written on my grandmother's death certificate. But I know better, and I knew better even then. Although the doctor explained what his scribbled words meant, his diagnosis made little sense to me, and it makes even less sense today. Had he wanted to call my Bubbeh's CVA the "terminal event" or some similar construct, I would have understood what was meant, but to tell me that the process I had been watching for eighteen years had ended in a named acute disease—well, it was illogical.

This is not simply a problem of semantics. The difference between CVA as a terminal event and CVA as a cause of death is the difference between a worldview that recognizes the inexorable tide of natural history and a worldview that believes it is within the province of science to wrestle against those forces that stabilize our environment and our very civilization. I am no Luddite—I glory in the magnificent benisons of modern scientific achievement. I ask only that we use our increasing knowledge with increasing wisdom. In the seventeenth and eighteenth centuries, the first of the early exponents of the experimental method, and therefore of science, spoke often of what they called the animal economy, and of the economy of nature in general. If I understand them correctly, they were speaking of that kind of natural law which exists to preserve the earth's environment and its living forms. That natural law, it seems to me, evolved by straightforward Darwinian principles of planetary survival, very much as did every species of plant or beast. For this to continue, mankind cannot afford to destroy the balance—the economy, if you will—by tinkering with one of its most essential elements, which is the constant renewal within individual species and the invigoration that accompanies it. For plants and animals, renewal requires that death precede it so that the weary may be replaced by the vigorous. This is what is meant by the cycles of nature. There is nothing pathological or sick about the sequence—in fact, it is the antithesis of sick. To call a natural process by the name of a disease is the first step in the attempt to cure it and thereby thwart it. To thwart it is the first step toward thwarting the continuation of exactly that which we try to preserve, which is, after all, the order and system of our universe.

And so, Bubbeh had to die, as you and I will one day have to die. Just as I had witnessed the decline of my grandmother's life force, I was present when it gave the first signal of its finality. It was early on an ordinary morning; Bubbeh and I were doing ordinary things. Having finished breakfast a few minutes before, I was still hunched over the sports section of the *Daily News* when I became aware that there was something very strange in the way Bubbeh was trying to wipe clean the surface of the kitchen table. Even though we had long since realized that such household tasks were beyond her, she had never quite given up trying, and seemed oblivious to the fact that one or another of us always repeated the work after she laboriously shuffled out of the room. But when I looked up from the tabloid, I saw that her wide circular strokes were even more ineffectual than usual. Her sweeping hand had become aimless, as though acting on its own with no plan or direction. The circles ceased to be circles and soon became mere languid, useless drags of the moist cloth that was barely held in her flaccid hand, adrift on the table without purpose or weight. Her face was turned straight ahead. She seemed to be looking at something outside the window behind my chair instead of at the table in front of her. Her unseeing eyes had the dullness of oblivion; her face was expressionless. Even the most impassive of faces betrays something, but I knew at that instant of absolute blankness that I had lost my grandmother. I shouted, "Bubbeh, Bubbeh!," but it made no difference. She was beyond hearing me. The cloth slipped from her hand and she crumpled soundlessly to the floor.

I bounded to her side and called her name again, but my shouting was as futile as my attempts to comprehend what was happening. Somehow, and I remember not a moment of it, I gathered her up and staggered to the room we shared. I laid her down in my bed. Her breathing was stertorous and loud. It blew in long, forceful blasts from only one corner of her mouth, and it flapped her cheek out like a buffeted wet sail each time she exhaled from that noisy bellows somewhere down deep in her throat. I can't recall which side it was, but one entire half of her face seemed toneless and flaccid. I rushed to the phone and called a doctor whose office was not far away. Then I contacted my aunt Rose at the Seventh Avenue dress factory where she worked. Rose got there before the doctor could free himself from a waiting room filled with early-morning patients, but we knew there was nothing he could do anyway. When he arrived, he told us that Bubbeh had suffered a stroke, and wouldn't live more than a few days.

She outsmarted the doctor, and hung on. We hung on with her, refusing to let go—it never occurred to anyone to do otherwise. Bubbeh remained in my bed, Aunt Rose occupied the double bed she had shared with her mother, and Harvey brought in his folding cot for me from the room in which he and my father slept. This left him adrift and he spent the next fourteen nights on the living room sofa.

Within forty-eight hours, we began to witness the most disheartening of the many cruelties by which life begins to desert its oldest friends—Bubbeh's worn-out immune system and her rusty old lungs were unable to withstand the blitzkrieg of microbes that now hurled itself against her. The immune system is the invisible force that allows us to respond to the assaults of potentially lethal enemies who are themselves invisible to the naked eye. Without our knowledge or conscious participation, the silent cells and molecules of immunity are ever adapting to the changing circumstances of daily life and its unseen terrors. Nature, our strongest shield and perforce our strongest enemy, has cloaked us and soaked us in them so that we may survive those constant encounters with the environment she has created (and is attempting to preserve), at the same time challenging each living thing to overcome the lurking perils of her constant testing. When we get older, the cloak becomes threadbare and the soak dries up—our immune system, like everything else, increasingly fails us.

The decline of the immune system has been a major focus of research by gerontologists. They have demonstrated defects not only in the elderly body's response to attack but even in the mechanisms of surveillance by which it recognizes its attackers. The enemy finds it easier to penetrate the perimeter by eluding immunity's aged watchmen; once inside, they overwhelm the weakened defenders. In my Bubbeh's case, the result was pneumonia.

William Osler was of two minds about pneumonia in the elderly. In the first of fourteen editions of *The Principles and Practice of Medicine*, he called it "the special enemy of old age," but elsewhere he stated something quite different: "Pneumonia may well be called the friend of the aged. Taken off by it in an acute, short, not often painful illness, the old escape those 'cold gradations of decay' that make the last stage of all so distressing."

I have no recollection of whether the doctor ordered penicillin to combat "the friend of the aged," but I doubt it. Selfishly perhaps, I didn't want Bubbeh to die, and neither did anyone else in our family. The doctor would have been much more realistic and a good deal wiser than we who refused to let go.

Bubbeh's immobile comatose state and the loss of her cough reflex prevented her from clearing the viscous mucilage of secretion that rattled in her windpipe with every breath. Harvey went off to the corner drugstore and discovered there a device that could be used to aspirate the increasingly purulent products that were rising from Bubbeh's lungs in a gurgly announcement of impending death. The instrument, made of two lengths of rubber tubing separated by a glass chamber, allowed him to suck out the phlegm each time it reaccumulated. It required him to put one rubber end of the device into Bubbeh's windpipe and the other into his own mouth. Even Aunt Rose couldn't bear to do it, and I could manage it only now and then, so it became Harvey's gift to his Bubbeh, or at least we thought of it as a gift.

By this means, and undoubtedly because of a change of heart by the Angel of Death himself (a figure of fancy to me, but a very serious reality to Old World believers), Bubbeh survived the pneumonia, and she even survived the stroke. Perhaps our tears and our prayers were more important than Harvey's mouth-operated suction device and the residual shreds of strength in that wheezing immune system of hers. Whatever it was, she slowly came out of her coma, regained a great deal of her speech and a small degree of mobility, and lived much as before for a few additional months, more for us than for herself. Finally, the days of her life ran out, and she succumbed to a second stroke in the early-morning hours of a chilly February Friday. In accordance with Jewish law, her body was in the ground by late afternoon of that day.

I have what some call a photographic memory. Although it sometimes deserts me when I am most in need of its collected transcripts, it has been for the most part a dependable ally in the record-keeping of my life. But there are some in my vast store of images I would rather lose. One of them is of an eighteen-year-old boy standing alone by the plain pine coffin of an ancient lady he can hardly recognize, even though he had tearfully kissed her unresponsive cheek scarcely twelve hours earlier. The object in the coffin looks so different from the Bubbeh it is said to be. It is contracted, and as white as candle wax. This corpse has shrunken away from life.

Nowadays, doctors are trained to think only about life and the diseases that threaten it. Even the autopsy pathologists who dissect corpses are looking for clues to cure, which are ultimately for the benefit of the living; in essence, what they do is to turn back the clock a few hours or a few days to a time when the heart still beat, in order to reconstruct the crime that stole their patient's life away.

Those among us who think most clearly about death are usually such as philosophers or poets, not physicians. Nevertheless, there have been a few doctors who understood that death and its aftermath are not beyond the limits of the human condition and are, therefore, worthy of a healer's attention.

Such a one was Thomas Browne, who lived in that extraordinary seventeenth century when the scientific method and inductive reasoning first began to affect the thinking of educated people and made them question the truths so dear to their fathers. In 1643, Browne published a small literary jewel of contemplation, *Religio Medici*, "The Religion of a Doctor," which he described as "a private exercise directed to myself." That little masterpiece is usually published together with a compilation on the lingering agonies of a dying man, entitled "A Letter to a Friend," in which the author writes: "He came to be almost half himself and left a great part behind him which he carried not to the grave." How often have I stood with families at a deathbed and witnessed their disbelief at this process unfolding its too-often-agonizing panorama before them. They question why it is different from their expectation and why seemingly they alone should have to endure what they conceive to be a uniqueness of suffering. It was that uniqueness I thought I had been forced to live through with Bubbeh's death and then later with the image of that alien corpse.

The force of life fills out our tissues with its pulsing vibrancy and puffs them up with the pride of being alive. Whether it departs with a bang, as it did for Irv Lipsiner, or a prolonged whimper, as it did for Bubbeh, it often leaves behind an object of shrunken unrealness. When Charles Lamb beheld the corpse of the popular English comedian R. W. Elliston, he was moved to write, "Bless me, how *little* you look. So shall we all look—kings, and kaisers—stripped for the last voyage." Browne himself wrote, "I am not so much afraid of death, as ashamed thereof; 'tis the very disgrace and ignominy of our natures, that in a moment can so disfigure us, that our nearest friends, wife, and children, stand afraid and start at us."

Thomas Browne's words, or Lamb's, might have reassured me at my grandmother's coffin. That day would surely have been a lot easier for me, and its memory less painful, had I but known that not only my own grandmother but indeed everyone becomes littler with death—when the human spirit departs, it takes with it the vital stuffing of life. Then, only the inanimate corpus remains, which is the least of all the things that make us human.

Reviewing those years just ended, I might also have recognized the commonality of death's experience in a sentence to be found a few pages earlier in Browne's book: "With what strife and pains we come into the world we know not, but 'tis commonly no easy matter to get out of it."

IV

Doors to Death of the Aged

MY GRANDMOTHER HAD chosen a way "to get out of it," to use Thomas Browne's expression, that is hardly unique. Stroke is the third most common cause of death in the developed countries of the world, as listed by the World Health Organization. More than one hundred and fifty thousand Americans die of it each year, representing approximately one-third of all those who suffer a stroke. Another third will be left with permanent severe disability. Only cardiac disease and cancer exceed its marauding power. Following a long period during which its incidence declined, a plateau has been reached in recent years: approximately 0.5–1.0 stroke per 1,000 population each year. But that figure represents the entire spectrum of our citizenry. As people age, their propensity to stroke naturally increases. There are no probability estimates for sedentary Jewish ladies who have kept themselves on a high-cholesterol kosher diet for almost a century, but we do know that of a random group of one thousand American or Western European men and women over the age of seventy-five, between twenty and thirty will suffer a stroke each year—among our eldest elders, the risk is some thirty times as great as it is for the rest of us.

Stroke is such a ubiquitous term that there is sometimes a little fuzziness about the way it is used. To a physician, a stroke is a deficit in neurologic function resulting from a decrease in blood flow through some specific artery supplying the brain. Further, the deficit must last longer than twenty-four hours for the episode to be called a stroke. Anything else is classified as a transient ischemic attack, or TIA. Although TIAs usually clear up within an hour, a few do last somewhat longer before their symptoms disappear.

If all of this has a familiar ring, there is good reason. It is basically the same mechanism by which the heart's deficit is produced when one of its arteries fails to deliver the required volume of blood. It is that universal mechanism of ischemia, the quenching of blood flow and the parching of tissues, that is so common a denominator in the killing off of cells in so many parts of the body. It carried off James McCarty, it carried off my Bubbeh, and, in one form or another, it will carry off most of us now living. It does its work by suffocating the tissues of its victim. The blood flow stops for much the same reason it does in the coronary arteries of the heart. The buildup of atheroma has reached the critical point at which a branch of one of the internal carotid arteries becomes completely occluded. The occlusion may be due to a completion of the atherosclerotic process in that branch itself, or it may occur because a bit of plaque has separated from the wall of a larger artery and been propelled as an embolus up into the brain, plugging an already-compromised vessel.

Or the stroke and its attendant ischemia may be caused by quite another manifestation of this vast syndrome of cerebrovascular disease, namely a hemorrhage into the substance of the brain, which in the elderly is almost always due to long-standing hypertension. Its wall already weakened by the long years of abnormally high pressure against it, the fragile atherosclerotic vessel finally gives way at some specific point, resulting in an outrush of blood into the surrounding brain tissue. Such an intracerebral bleed carries a mortality rate of twice the 20 percent that is usually estimated for occlusive strokes. Hemorrhage accounts for approximately 25 percent of strokes, and vascular occlusion for the rest.

It takes a great deal of energy to keep the brain's engine functioning efficiently. Almost all of that energy is derived by the tissue's ability to break down glucose into its component parts of carbon dioxide and water, a biochemical process that requires a high level of oxygen. The brain does not have the capacity to keep any glucose in reserve; it depends on a constant immediate supply being brought to it by the coursing arterial blood. Obviously, the same is true of the oxygen. It takes only a few minutes for the ischemic brain to run out of both before it suffocates. Neurons are extremely sensitive to ischemia; irreversible destructive changes begin within fifteen to thirty minutes of the onset of the deprivation. After no more than an hour of the beginning of ischemia, infarction of significant segments of brain tissue becomes inevitable.

The symptoms caused by the cell destruction vary, depending upon which vessel is occluded. Although at least half a dozen branches of the internal carotid are particularly susceptible to obstruction, most frequently involved in ischemic stroke is one of the paired middle cerebral arteries. The middle cerebral (MCA) supplies blood to most of the lateral surface of the cerebral hemisphere and some of the centers that lie deep beneath the cortex. The MCA feeds the major sensory and motor areas of the cortex—areas that are involved in hand and eye movement, and also the specialized sensory tissue for hearing. It nourishes the region involved in what are called the "higher mental functions," such as perception, organized thought, voluntary movement, and the integrated coordination of all these abilities. On the dominant side of the brain (the right side for lefties and the left for the other 85 percent of us), the MCA supplies the sensory and motor areas for language. This particular bit of geographic distribution explains why so many stroke victims lose their powers to express and comprehend spoken and written language.

Many MCA strokes are caused not by actual occlusion at the site but by bits of material broken off from the crusted atheroma in the main internal carotid artery, or from the heart itself in the form of small bits of old organized clot. The released particle then becomes an embolus. Here we encounter another of those terms contributed by Rudolf Virchow, from the Greek *embolos*, signifying a wedge or plug, which in turn is derived from two words meaning "to cast or throw in." Literally, then, a plug has been thrown into the artery and will be propelled by the bloodstream until it wedges itself into a narrowed portion of the vessel, which it will completely block off. In the more common cases where the plugging is not caused by an embolus, it results from the final completion of atheroma buildup. In either situation, the tissue supplied by the vessel instantly loses its source of oxygen and glucose, and within a few minutes becomes wounded enough to cause symptoms. If the blockage is not relieved rapidly, the area of brain dies by infarction.

If one were to name the universal factor in all death, whether cellular or planetary, it would certainly be loss of oxygen. Dr. Milton Helpern, who was for twenty years the Chief Medical Examiner of New York City, is said to have stated it quite clearly in a single sentence: "Death may be due to a wide variety of diseases and disorders, but in every case the underlying physiological cause is a breakdown in the body's oxygen cycle." Simplistic though it may sound to a sophisticated biochemist, this pronouncement is all-encompassing.

Many strokes are so small that there are few or no immediate significant symptoms to indicate what has taken place. But with time, such little strokes accumulate, and the evidence of gradual deterioration becomes clear to even the most casual observer. Walter Alvarez, the great Chicago clinician of a generation ago, once quoted "a wise old lady" who said to him, "Death keeps taking little bits of me." As his clinical description so clearly states:

> She saw that with each attack of dizziness or fainting or confusion she became a little older, a little weaker, and a little more tired; her step became more hesitant, her memory less trustworthy, her handwriting less legible, and her interest in life less keen. She knew that for 10 years or more, she had been moving step by step towards the grave.

Of those so betrayed by their cerebral circulation, William Osler is reported to have said, "These people take as long to die as they did to grow up."

Almost 10 percent of elderly people diagnosed with dementia owe their situation to a series of such small strokes, a concept popularized by Alvarez in 1946, after observing it in his own father. Now called multi-infarct dementia, the process is characterized by an irregular series of abrupt little worsenings. Interestingly, this form of cerebral arteriosclerosis was first described by Alois Alzheimer in 1899, eight years before he introduced the quite different type of intellectual decline that we now call by his name.

The subtle process of infarcting brain may go on and on, accumulating irregular stepwise degenerations in cerebral function for as long as a decade or more, until a major stroke or some other lethal process intervenes to bring abrupt fulfillment of the slow progression's ultimate purpose.

Major infarction by MCA stroke results in sensory loss and weakness that are most prominent in the part of the face and in the extremities opposite to the side of the brain where the stroke has occurred; such infarction causes as well a condition called aphasia—the loss of power of expression—although comprehension tends to remain reasonably well preserved. Occlusion of other vessels produces a whole range of symptoms, determined not only by the area served by the vessel but also by the amount of nutrition brought in by a collateral blood supply that may be available from nearby unscathed vessels. Language and visual disturbances, paralysis, sensory losses, difficulties in balance—all of these are the more common manifestations of stroke.

Large strokes often result in coma. If the stroke is extensive enough or if further complications ensue, such as decreased blood pressure or cardiac output due to failure or arrhythmia, recovery is prevented and the area of ischemia may actually increase. If it becomes large enough, the brain tissue begins to swell. Being compressed in the unyielding confines of the skull, a swollen brain is further damaged by being pushed up against its covering membranes and bony encasement, and part of it may actually be forced down through a fold in those membranes that separates the "higher" brain from the "lower," or brain stem—the part that thinks from the part that is involved with more automatic mechanisms such as cardiac and respiratory control, digestive and bladder function, and a group of others. When this happens, the pressure causes so much damage to the brain stem's centers controlling the heart and breathing that death follows soon thereafter, from either arrhythmia or cardiac and respiratory failure.

Collapse of vital function is only part of the array of possible mechanisms by which strokes kill approximately 20 percent of their victims, or even more when the cause is hypertensive hemorrhage. If brain damage is massive enough, all kinds of normal controls go awry. Preexisting diabetes sometimes goes so much out of control that blood acidity rises enough to be incompatible with life; the functioning of the lungs is sometimes impaired by paralysis of the muscles of the chest wall; the blood pressure may become elevated to dangerous levels—these are among the most common lethal complications of major strokes.

And then there is my Bubbeh's way—pneumonia. More than any other organ system excepting skin, the lungs of elderly people are subjected to every insult our polluted environment is capable of inflicting. Whether elasticity is lost for that reason or simply by the ordinary process of aging, the passage of time results in a decreased ability to inflate or deflate completely. Mechanisms for clearing mucus are weakened, and the already-narrowed airways are more prone to become filled with debris-laden material. The situation is worsened by an inability to maintain proper humidity and temperature in the finer bronchial branches. These strictly physical debilities are complicated by a depressed production of local antibodies as part of the old person's generally lessened immune response.

The microbes of pneumonia lie in wait for the appearance of any added insult that might inhibit further the already-damaged defenses of the aged. Coma is their perfect ally. It takes away every conscious way of resisting their predations, and even destroys so basic a safety device as the cough reflex. Any bit of regurgitation or foreign matter that under ordinary circumstances would be forcefully ejected at the first sign of its invasion of the airway now becomes the vehicle on which the germs ride triumphantly into the respiratory tissues. The microscopic air sacs called alveoli then swell and are destroyed by inflammation. As a result, proper exchange of gases is prevented, and blood oxygen diminishes while carbon dioxide may build up until vital functions can no longer be sustained. When oxygen levels drop below a critical point, the brain manifests it by further cell death, and the heart by fibrillation or arrest. Pneumonia triumphs.

Pneumonia's blitzkrieg has yet another way to kill—its putrid headquarters in the lung serves as a focus from which the murderous organisms can enter the bloodstream and be dispersed into every organ of the body. Called sepsis or septicemia by doctors and blood poisoning by the rest of the world, this process sets off a series of physiologic events that results in collapse of the integrity of heart, lungs, blood vessels, kidneys and liver, with an ultimate drastic drop in blood pressure to shock levels, followed by death.

In sepsis, even the most powerful antibiotics are often no match for the microbes' overwhelming assault.

Whether the terminal event is pneumonia, heart failure, or the acidosis of uncontrollable diabetes, the salient fact about stroke is that it is always to be found in the company of its friends—the ubiquitous corps of killers of the old. A stroke is simply one part of the wide spectrum of end-stage cerebrovascular disease, which, while it may be hastened by self-abuse, can never be stopped in its determined course. Henry Gardiner, who compiled my 1845 edition of Thomas Browne's writings, has bound into its appendix a long quotation from Francis Quarles, a seventeenth-century literary figure, who properly said: "It lies in the power of man, either permissively to hasten, or actively to shorten, but not to lengthen or extend the limits of his natural life." And then in a sublime bit of wisdom, Quarles added, "He only (if any) hath the art to lengthen out his taper, that puts it to the best advantage." There is no way to deter old age from its grim duty, but a life of accomplishment makes up in quality for what it cannot add in quantity.

Many doctors, especially those who spend much of their time in laboratories, share with statisticians the disbelief in the necessity of death from old age. Reading my account of my Bubbeh's last days, they will by now undoubtedly have pointed out that pneumonia and other infections become, after all, the second most common identifiable cause of death once people reach the very great age of eighty-five, and atherosclerosis is the first. My grandmother had both, and therefore, they may claim, her mode of death supports their worldview and argues for vigorous intervention to treat the named pathologies in order to prolong life. To me, this is more sophistry than science.

I grant these doctors their perspective, but there is plenty of evidence that life does have its natural, inherent limits. When those limits are reached, the taper of life, even in the absence of any specific disease or accident, simply sputters out.

Fortunately, most bedside doctors who restrict their practice to the care of the aged have come to understand this. The geriatricians are to be applauded for the great contributions they have already made to elucidating the pathologies afflicting those who are slowly being overwhelmed by the defects in their waning senescent powers, but even more so do they deserve our admiration for the compassion they bring to their work. I recently discussed this with my school's professor of geriatric medicine, Dr. Leo Cooney, who later summarized his viewpoint in two pithy paragraphs of a letter:

> Most geriatricians are at the forefront of those who believe in withholding vigorous interventions designed simply to prolong life. It is geriatricians who are constantly challenging nephrologists [kidney specialists] who dialyze very old people, pulmonologists [lung specialists] who intubate people with no quality of life, and even surgeons who seem unable to withhold their scalpels from patients for whom peritonitis would be a merciful mode of death.
>
> We wish to improve the quality of life for older individuals, not to prolong its duration. Thus, we would like to see that older people are independent and lead a dignified life for as long as possible. We work to decrease incontinence, manage confusion, and help families deal with devastating illnesses like Alzheimer's.

Basically, geriatricians can be viewed as the primary care doctors for elderly people, this generation's solution to the problem of the absence of the old family doctor who knew his patients as well as he knew their diseases. If a geriatrician is a specialist, his specialty is the entire elderly person. In late 1992, there were only 4,084 certified geriatricians in the United States; at the same time, there were some 17,000 heart specialists.

One might question certain portions of my evidence for saying that the natural limits of an individual's life permit little tampering. There are in fact some very elaborate studies that have been conducted on aging people who have remained well. In those investigations, age-specific changes in function were evaluated in men and women who had no disease process that could be expected to affect that function. The results are as I have described them—the aging process goes on, regardless of anything else that may be happening. Aging may be said to be both independent and codependent, in the sense that it certainly contributes to disease and may in turn be accelerated by it. But disease or no disease, the body continues to get older.

My disagreement with the viewpoint of many of those laboratory researchers who study the physiology of aging centers around the philosophy of treatment. When it is possible to identify a disease by giving it a name, its ravages become the subject of treatment, with the potential aim of cure. And that, after all, is the real reason a modern scientific doctor becomes a specialist. No matter his stated interest in relieving human suffering and no matter the sincerity of his efforts, the average specialist physician does what he does because he is absorbed by the riddle of disease and longs to conquer it by solving each puzzlement it presents to his inquisitive mind, whether he is a researcher or a clinician. At each end of life, the pediatric and geriatric age groups, patients are fortunate to be guided by one of today's equivalents of the family doctor.

The diagnosis of disease and the quest for overcoming it with his intellect are the challenges that motivate every specialist who is any good at what he does. He is fascinated with pathology. When faced by the certainty of his own impotence to treat it, the would-be healer too often turns away. If a riddle is by its nature insoluble, it cannot long hold the interest of any but a tiny fraction of the doctors who treat specific organ systems and disease categories. Old age is as insoluble as it is inevitable. By giving scientific names of treatable diseases to its manifestations, too many of the specialists from whom the elderly seek care retain their riddle and their fascination. They also believe they give patients some kind of hope, though in the end the hope must always prove to be unjustified. These days (if I may steal a term from the jargon of the contemporary rialto), it is not politically correct to admit that some people die of old age.

Can there be any doubt that the intrinsic physical processes associated with aging inevitably cause an individual to become progressively more vulnerable to mortality? Can there by any doubt that every year we grow less able to marshal the forces required to fight off the lethal dangers that lurk constantly around us? Can there be any doubt that this growing inability is the result of gradually incremental debilitation in the powers of our tissues and organs? Can there be any doubt that the debilitation is due to a general running down of normal structure and function? Can there be any doubt that a general running down, whether in a motor or a man, will eventually lead to nonfunction? Can there be any doubt that Thomas Jefferson knew what he was talking about?

The understanding expressed by Jefferson is in fact millennia-old. In the oldest extant medical book of China, or anywhere else—the *Huang Ti Nei Ching Su Wen* (*The Yellow Emperor's Classic of Internal Medicine*), written some 3,500 years ago—the mythical emperor is being instructed about old age by the learned physician Chi Po, who tells him:

> When a man grows old his bones become dry and brittle like straw [osteoporosis], his flesh sags and there is much air within his thorax [emphysema], and pains within his stomach [chronic indigestion]; there is an uncomfortable feeling within his heart [angina or the fluttering of a chronic arrhythmia], the nape of his neck and the top of his shoulders are contracted, his body burns with fever [frequent urinary-tract infections], his bones are stripped and laid bare of flesh [loss of lean muscle mass], and his eyes bulge and sag. When then the pulse of the liver [right heart failure] can be seen but the eye can no longer recognize a seam [cataracts], death will strike. The limit of a man's life can be perceived when a man can no longer overcome his diseases; then his time of death has arrived.

The major question is not *whether* aging leads to debility, the inability to overcome disease, and then death, but *why* individuals age in the first place. The Preacher of *Ecclesiastes* was among the first in the Western tradition to point out that "To every thing there is a season, and a time to every purpose under the heaven: A time to be born, and a time to die," but the theme is so commonplace as to echo through our literature in every era. Before the Preacher, Homer had written, "The race of men is like the race of leaves. As one generation flourishes, another decays." And there are good reasons that one generation must give way to the next, as made clear in another of the letters Jefferson wrote to the equally venerable John Adams near the end of his life: "There is a ripeness of time for death, regarding others as well as ourselves, when it is reasonable we should drop off, and make room for another growth. When we have lived our generation out, we should not wish to encroach on another."

If it is the way of nature that we not "encroach on another" (and simple observation confirms that it is), then nature must of necessity provide some means of certainty that we, like Homer's leaves, progressively attain a stage at which we "drop off, and make room for another growth," as gentleman farmer Jefferson put it. Scientists of every stamp have attempted to identify the mechanism by which living things do this, and we still don't know for certain what it is.

Basically, there are two distinct lines of reasoning to explain the aging process. One emphasizes the continued progressive damage done to cells and organs by the commonplace process of carrying out their normal functions in the ordinary environment of everyday life. This is often called the "wear and tear" theory. The other suggests that aging is due to the existence of a genetically predetermined life span that controls not only the longevity of individual cells but of organs and entire organisms, like ourselves, as well. In descriptions of this latter thesis, the image is often invoked of a "genetic tape" that begins to run at the instant of conception and plays out a sequential program that preordains not only the hour of death (at least in the metaphoric sense) but even the hour at which the death-dealing notes begin to he heard. Carried to its most specific implication, this theory might mean, for example, that the day or week of a cancer's first cell division has already been determined at the time the same event is happening in the just-fertilized egg.

As used by the proponents of the "wear and tear" theory, the word *environment* may refer to the environment of this planet or the environment within and around the cell itself. It may be that such factors as background irradiation (both solar and industrial), pollutants, microbes, and toxins in the atmosphere slowly result in damage that changes the nature of the genetic information transmitted by cells to their offspring. It may even be that the environment plays no part—the misinformation may result from random errors in transmission. Either way, the accumulated alterations in DNA might then cause the errors in a cell's function that lead to its death and those obvious changes in the whole organism that manifest themselves as aging. This process of frank cellular death is called by some the "error catastrophe."

Some of the environmental hazards originate within our tissues and inside the cell. I have already described the constant bombardment that affects the basic nature of molecules, but there are other mechanisms as well. In order to continue in vibrant health, cells must efficiently break down the toxic products of their own metabolism. If there is any degree of escape from this mechanism, the harmful by-products may accumulate and affect not only function but the DNA itself; whether its cause is the environment, random errors in transmission, or toxic products of metabolism, the development of errors in the DNA is thought by many to be a major factor in the aging process.

Although we should not take the fright literature of New Age doom describers too seriously, there is no doubt that some of their shibboleths, such as aldehydes and free radicals of oxygen, demand attention because they may play a role in the damaging and aging of protoplasm if they are not properly degraded into less hazardous substances. A free radical is a molecule whose outer orbit contains an odd number of electrons. Such structures are extremely reactive, because stability can be acquired only by gaining an electron or losing the one that is unpaired. The extreme reactivity of free radicals has made them either the culprit or the hero of numerous biological theories, ranging from the very origins of life on this planet all the way along the spectrum to mechanisms of aging. Some of our more activist would-be extenders of life are convinced that an extra load of beta carotene or vitamin E or C in the diet will rescue our tissues from oxidation by free radicals. Unfortunately, there is as yet no definitive evidence to prove that they are correct.

The other of the two major theories of aging is the proposition that the entire process is predetermined by genetic factors. In this formulation, there exists within each living thing a genetic program whose function is to progressively shut down the physiological processes of normal life and eventually of any life at all. Among humans, different people do it in different ways, or at least its most prominent features vary in each of us. This allows for such separate phenomena as loss of immunity, wrinkling of the skin, the growth of malignancies, the onset of dementia, decreased elasticity in blood vessels, and many other events of senescence.

The genetic theory was given a huge boost almost thirty years ago when Dr. Leonard Hayflick showed that human cells cultured in the laboratory begin to slowly stop dividing after a while. In time, they quit altogether, and die. The maximum number of cell divisions was found always to be finite, and to be about fifty. The studies were conducted on a ubiquitous type of cell called the fibroblast, which makes up the basic structural framework of all tissues of the body, and the findings may be extrapolated to other cells as well. The seemingly endless capacity of the cancer cell to reproduce, of course, escapes the orderly finiteness of normal existence.

Such studies as Hayflick's help to explain why each species exhibits a characteristic life span and why individuals within species tend to have life spans that correlate well with those of their parents—the best assurance of longevity is to choose the right mother and father.

A plethora of specific aging factors have pushed their way onto the scene of science, and my guess is that virtually all of them have some degree of validity. In other words, aging is very likely the result of all of them in combination, with the importance of individual components varying for each of us. Some of the factors are common to all living things. Among them are the changes that occur in molecules and organelles. Changes that occur in cells, tissues, and organs may be specific to a single species, like those that take place in the whole plant or animal. The evidence, as Dr. Hayflick puts it, "is strongly persuasive that those attributes of biological instability that by common belief are considered to be age changes have a multiplicity of causes."

Some of the biological phenomena have already been described, such as the genetic program itself, the generating of free radicals, the instability of molecules, finite cell life, and accumulated genetic and metabolic errors. There are several other possible components that have found vigorous champions in the halls of science. Lipofuscin, for example, is thought by some researchers to be more than simply a bland product of intracellular breakdown that harmlessly discolors aging organs; they believe that its accumulation is lethal. Others place great emphasis on hormonal changes mediated through the nervous system; there are proponents of the theory that among the changes occurring in the immune system, one of the most fundamental is its decreasing ability to recognize its own host's tissues, with the result that the degenerative diseases of the elderly are brought about by the body's rejection of some of the very tissues of which it is made.

Yet another theory holds that the molecules in the structural tissue called collagen become cross-linked to one another. The aggregation of such linkages impedes the flow of nutrients and wastes while at the same time decreasing the space needed for vital processes to occur. Among its other effects, cross-linking may damage DNA, which in turn brings about mutations or cell death. And there is a relatively new theory that physiological systems and perhaps also the anatomic changes that may accompany them become less complex with age, and therefore less efficient; the decreased complexity may be the result of other more basic processes, perhaps including some of those already described.

Recently, moreover, there has been increasing interest in a phenomenon widespread among species, which seems to be a programmed form of cell death. This process, which researchers have named apoptosis (from the Greek, meaning "a falling away from"), is initiated by the activity of a protein called the myc gene, which starts a powerful series of genetic reactions under specific abnormal circumstances. For example, when nutrients are removed from certain types of cells growing in culture, the myc gene begins a process by which the cell undergoes an event that resembles an implosion, which in the course of about twenty-five minutes destroys it. It quite literally "falls away from" life. Such programmed death is important to the development of the mature organism, because it is a means whereby certain cells that are no longer useful in the developmental process can be replaced by those that belong to the next phase. Examples have also been found of apoptosis in fully mature individuals, triggered by various events in the environment of the affected cells.

Because apoptosis is a situation in which cell death occurs as a direct consequence of gene expression, it is tempting to ruminate over the possibility that the myc protein or something very like it can function as a "death gene." This gene-directed death may be instigated by a variety of environmental and physiological factors, and it seems to provide a concordance between some of the different theories described in the foregoing paragraphs. This research pathway is made all the more promising by the demonstration of binding between the myc protein and another structure given the name max protein. When they bind, the cell is instructed in some not-yet-understood way to do one of three things—mature, divide, or self-destruct by apoptosis. Depending on how it expresses itself, therefore, the myc gene obviously plays a major role in development, in growth regulation, and finally in a programmed form of death. The implications of these new discoveries are obviously incalculable at the moment, not only for the understanding of normal processes but for pathological ones as well, particularly cancer.

The proponents of collegial compromise are exploring yet other avenues that may lead to clarification of seemingly disparate viewpoints. For example, the immune changes of senescence may be the result of hormonal influences determined by neurological events that are in turn genetic—or vice versa. There is no lack of theories, no lack of champions, and no lack of concordances among concepts. What emerges from all the experimental data and the speculations they provoke is the inevitability of aging, and therefore of life's finiteness.

And what about those federally funded lists of formally named pathologies by which old people are supposed to die? Every group of lethal diseases of the elderly consists predominantly of the usual suspects. Of hundreds of known diseases and their predisposing characteristics, some 85 percent of our aging population will succumb to the complications of one of only seven major entities: atherosclerosis, hypertension, adult-onset diabetes, obesity, mental depressing states such as Alzheimer's and other dementias, cancer, and decreased resistance to infection. Many of those elderly who die will have several of them. And not only that; the personnel of any large hospital's intensive care unit can confirm the everyday observation that terminally ill people are not infrequently victims of all seven. These seven make up the posse that hunts down and kills the elderly among us. For the vast majority of those of us who live beyond middle age, they are the horsemen of death.

Autopsies are not as popular as they were a few decades ago. Given the meticulous accuracy with which diagnoses can nowadays be made before death, autopsy has become in the eyes of many bedside physicians a redundant exercise in academic pathology. Far fewer people die of a wrong diagnosis than did in an earlier era—the enormous majority succumb because of our inability to change the course of an accurately pinpointed disease. In the last decade or more, my own hospital's so-called autopsy percentage has dwindled to a level that hovers around 20 percent, when for many years before then it stood consistently at well above twice that figure. The national rate is now about 13 percent.

During the heyday of the autopsy, I was able to obtain postmortem permission from the families of all but very few of my own patients who succumbed. I don't try as hard these days, but when I do, I still make a point of being present to review the findings with the pathologist as he or she reveals them. After six years of residency training and thirty years of practice, I have witnessed a rather large number of autopsies. The wide extent of atherosclerosis and atrophy to be found in the bodies of old people is a commonplace, seemingly unworthy of comment when the dissector is seeking out the several sites to which a cancer may have spread or an infection lodged. Peering assiduously at tissues and into the insides of organs, both the dissector and the surgeon tend to ignore the familiar panorama of aging that gradually reveals itself with every added stroke of the knife. Remarking on it is as unusual as a motorist commenting on the leaflessness of a winter landscape when looking for the right street address—it's just there, and that's all there is to it.

And yet, when the autopsy report arrives in the surgeon's—my—mailbox a few weeks later, I have often been astonished at the very advanced state of the barely noticed biological debris through which the pathologist and I have so recently made our way. In the detailed analysis of his findings, he has meticulously entered every one of his discovered divergences from normal health. As I read his summary, they all spring back to memory and take their place alongside the major clues we were so single-mindedly tracing. Only when this begins to happen do I fully appreciate the entire setting in which my patient has died.

Some of the autopsy findings have nothing to do with the circumstances of death. They are simply the results of the selfsame aging process out of which one or two particular kinds of pathology emerged to kill the patient. Such non-lethal findings may not directly contribute to the death, but they provide the background against which it occurs.

Recently, I sought out the help of a colleague at the Yale–New

Haven Hospital. Dr. G. J. Walker Smith is the director of the autopsy service, an astute veteran of that marbled chamber in which the doctors of the dead strive mightily to answer the question framed more than two hundred years ago by the founder of their somber specialty, the Paduan anatomist Giovanni Battista Morgagni: *Ubi est morbus?* "Where is the disease?" Together, pathologist and newly deceased patient undertake the obligation promised by the time-honored avowal that stares down on them from plaques on the walls of hundreds of autopsy rooms all over the world: *Hic est locus ubi mors gaudet succurso vitae*—"This is the place where death rejoices to come to the aid of life."

The autopsy room is Walker Smith's domain, just as the operating room is mine. When I told him I was interested in confirming my long-standing impressions by seeing a few final reports on patients who had died in advanced age, he did me one better—he himself became interested, and before long was just as taken with the project as I. He found twenty-three records of patients whose studies had been done before today's scarcity. Together, we reviewed the findings in twelve men and eleven women eighty-four years of age and older who had died in the sixteen-month period between December 1970 and April 1972. Their average age was eighty-eight, and the oldest was ninety-five.

Although there were variations in the distribution of such pathologies as atherosclerosis and microscopic deterioration of the central nervous system, when viewed en masse, there was nevertheless a sameness about the findings that was starkly impressive to both of us.

An individual's specific kind of death seems to depend upon the order in which his tissues become involved in the process of degradation. The one common thread among the twenty-three patients, at least as it is reflected in the staccato multisyllables of a pathologist's unique approach to obituary, was the loss of vitality that comes with starvation and suffocation—as the arteries narrow, so does the margin between life and death. There is less nutriment, there is less oxygen, and there is less resiliency after insult. Everything rusts and crusts until life is finally extinguished. What we call a terminal stroke, or a myocardial infarction, or sepsis is simply a choice made by physicochemical factors we do not yet comprehend, the purpose of which is to bring down the curtain on a performance already much closer to its conclusion than may have been realized, even in an old person who has till then appeared vibrantly healthy.

An octogenarian who dies of myocardial infarction is not simply a weather-beaten senior citizen with heart disease—he is the victim of an insidious progression that involves all of him, and that progression is called aging. The infarction is only one of its manifestations, which in his case has beaten out the rest, though any of the others may be ready to snap him up should some bright young doctor manage to rescue him in a cardiac intensive care unit. Seven of Walker Smith's oldsters officially died of myocardial infarction; four others had strokes as part of their terminus; eight died of infection, including three who disappeared into eternity arm in arm with the old man's friend, pneumonia; there were three far-advanced cancers in the group, although the final episode for one was pneumonia and for another was stroke. The single most striking observation was also the one most expected: Every one of these twenty-three people had advanced atheromatous disease in the vessels of the heart or the brain, and almost all had it in both, even if they exhibited no symptoms that required treatment until the terminal event. One or the other of these vital engines was close to quits in every one of the old persons studied.

Another finding that elicited no surprise was the frequency of nameable disease in the other organs of any individual, which played no part in the patient's death. In the pathologists' reports, such diseases are designated "incidental." Thus, in addition to the three patients whose malignancies killed them, another three were found to harbor unsuspected "incidental" cancers—in the lung, prostate, and breast; two women and one man had the ballooning of the aorta or other large abdominal vessels called an aneurysm, caused by atherosclerotic weakening; eleven of the twenty whose brains were microscopically studied were found to have old infarcts, even though only one had a known history of stroke; fourteen were found to have major atherosclerotic changes in the arteries of the kidneys; several had active urinary-tract infections; and a man who died of extensive stomach cancer had gangrene of his leg.

It is well known that very old people die of diseases they might easily have conquered had they been somewhat younger, but the extent to which this happens in the case of perfectly straightforward sickness is surprising: One of the people in our study died of a ruptured appendix, two of infections following gallbladder or bile-duct surgery, one from the complications of a perforated peptic ulcer, and another of diverticulitis. Every one of these diseases is an infection. Infection is exceeded only by atherosclerosis as the most frequent cause of death in people eighty-five or older. Two additional patients died of hemorrhage—one in a duodenal ulcer and one as the result of a fractured pelvis. Having been in the midst of a very active surgical practice during the period when these autopsies were done, I can attest to the probability that none of the seven individuals being treated in this university hospital would have succumbed had they been in their mid-fifties.

Only two of Walker Smith's twenty-three patients escaped significant destruction of brain tissue. One of them proved, in fact, to be remarkably resistant in general to atherosclerosis, at least of the brain and heart. The degree of calcification in this eighty-nine-year-old man's coronary arteries was only moderate and he had sustained "less cerebral atrophy than might be expected in a brain of this age," to quote the autopsy report. But he had taken it in the kidneys, which were not only the site of a chronic infection (called pyelonephritis) that constantly seeded his urinary tract with intestinal bacteria but were also victimized by destruction of their tiny arterial branches and filtering units, as well as by marked scarring. Yet it was not his chronic kidney disease that did this fellow in—he succumbed to a malignancy called multiple myeloma, complicated by pneumonia. And so, like every other one of these twenty-three very old men and women, this fellow was carried off by several of the seven horsemen.

The other escapee from the ravages of cerebral senescence was an eighty-seven-year-old professor of Latin and former Yale dean. Seemingly spry and well (and without clinical evidence of heart disease), he was discovered at autopsy to have been actually within a hair of myocardial infarction, with the interesting coupling of "severe [atherosclerotic] involvement of the coronary arteries and minimal involvement of the cerebral vessels." His coronaries, in fact, were described as "pipe-stem," with one of them being completely occluded. The heart had undergone a brownish discoloration due to atrophy; the kidneys, too, looked their age. The professor had been awakened from sleep one cold December night by the sudden onset of severe abdominal pain. The diagnosis of perforated peptic ulcer was made in the emergency room and confirmed at autopsy four days later, after his tired immune system and barely nourished heart proved insufficient to protect him from the peritonitis that ensued. And so the professor's relatively unscathed brain was of no avail to him when his life was challenged elsewhere.

The lesson taught by the twenty-three case histories is simply confirmation of the lesson that daily experience teaches. Whether it is the anarchy of disordered biochemistry or the direct result of its opposite—a carefully orchestrated genetic ride to death—we die of old age because we have been worn and torn and programmed to cave in. The very old do not succumb to disease—they implode their way into eternity.

Since there are so few pathways to an old man's grave, and since there is such an intermingling of their basic paving stones, it is reasonable to wonder why the development of one of them brings with it such a high risk of harboring the others. Is it perhaps that all of these pathologies may share a common cause that becomes more active as we grow older? This consideration has, of course, been incorporated into the various theories of aging. One of the theories, for example, proposes that the process by which we develop and grow is part of a metabolic pattern controlled by an inner part of the brain called the hypothalamus, which can regulate hormone activity. This mechanism, beginning when life itself begins, allows the body to adapt to its external environment. The progression of these adaptations necessarily leads, as though following a schedule, to development, maturity, and then aging. If there is truth to this neuroendocrine thesis of aging, the occurrence of the diseases of the elderly is the price an organism pays for its lifelong ability to adapt to its surroundings and to changes occurring in its own tissues.

The entire process unfolds as though part of a master plan, a grand strategy that oversees an organism's development from early embryonic stages to the instant of mortality, or at least to the anarchy that immediately precedes it. In this, the theorists of physiology are at one with the bereavement counselors who point out the value of the maxim that death is part of life.

Considerations of this sort echo, albeit in a more somber vein, a few sentences from a passage in the appendix of my volume of Thomas Browne. In a book entitled *Merchant and Friar*, the nineteenth century historian Sir A. Palgrave wrote: "Coeval with the first pulsation, when the fibres quiver, and the organs quicken into vitality, is the germ of death. Before our members are fashioned, is the narrow grave dug, in which they are to be entombed." Dying begins with the first act of life.

These are possibilities that give rise to speculations of major significance in decision-making about our own lives. When an elderly man is offered the possibility of cancer palliation or even cure, providing that he is willing to endure debilitating chemotherapy or radical surgery, what should be his response? Will he suffer through the treatment, only to die of his ongoing cerebrovascular atherosclerosis the following year? After all, the cerebrovascular disease is likely the result of the same process that so decreased his immunity to malignant growth that he developed the cancer that is trying to kill him. But then again, different manifestations of the aging process proceed at different rates, so it may be somewhat longer than he anticipates before his stroke exerts its claim. Such possible eventualities can be estimated only by evaluating the present state of his nonmalignant process, such as the degree of his hypertension and the status of his heart disease. These are the kinds of considerations that should go into every clinical decision involving older people, and wise physicians have always made careful use of them. Wise patients should do the same.

Whether the result of wear, tear, and exhaustion of resources or whether genetically programmed, all life has a finite span and each species has its own particular longevity. For human beings, this would appear to be approximately 100 to 110 years. This means that even were it possible to prevent or cure every disease that carries people off before the ravages of senescence do, virtually no one would live beyond a century or a bit more. Although the psalmist sings that "The days of our years are three score and ten," it seems not to be remembered that Isaiah was a better prophet, or at least a better observer, proclaiming to all who would but listen that "the child shall die an hundred years old." He is here speaking of the New Jerusalem, where there will presumably be no infant mortality and no disease: "There shall be no more thence an infant of days, nor an old man that hath not filled his days." Were we to heed Isaiah's warning and eschew every bit of McCarty-like behavior, solve the problems of poverty, and love our neighbor, who knows how close we might come to making a prophet of the prophet? Medical science and improved living conditions have already brought us a long way. Western society has in less than a hundred years more than doubled a child's life expectancy at birth. We have changed the face of death. In the modern demographic pattern, the great majority of us now reach at least the first decade of old age, and we are fated to die of one of its ravages.

Though biomedical science has vastly increased mankind's *average* life expectancy, the *maximum* has not changed in verifiable recorded history. In developed countries, only one in ten thousand people lives beyond the age of one hundred. Whenever it has been possible to examine critically the claims of supposed record-breakers, they have not been substantiated. The highest age thus far solidly confirmed is 114. Interestingly, that figure comes from Japan, whose citizens live longer than those of any other country, with an average life expectancy of 82.5 years for women and 76.2 for men. The comparable figures for white Americans are 78.6 and 71.6, respectively. Even the home-cultured yogurt of the Caucasus cannot vanquish nature.

There is plenty of other evidence to support the thesis of a species-determined limit to life span. Among the most obvious clues is the great variability in the maximum attainable age between differing animal groups, existing coincident with the highly specific longevity of each individual species. Another suggestive biological observation is the average number of offspring of any animal form, which proves to be inversely related to the form's maximum life span. An animal like man, needing not only a considerable gestation period but an inordinately long time before its young are biologically independent, requires a prolonged reproductive life span to ensure survival of the species, and that is exactly what we have been given. Humans are the longest-living mammals.

If the processes of aging are, within relatively narrow limits, resistant to any but certain well-known changes in personal habits, why do we persist in heretofore-vain attempts to live beyond the possible? Why cannot we reconcile ourselves to the immutable pattern of nature? Although recent decades have seen our concern with our bodies and their longevity reach a fever pitch unknown to previous generations, these kinds of hopeful seekings have always motivated at least some members of those societies that have left records of their existence. As early as the days of ancient Egypt, there is evidence of attempts by elders to prolong their lives—the Ebers Papyrus of more than 3,500 years ago contains a prescription for restoring an old man to youth.

Even as science was beginning to light the dawn of a new kind of medicine in the seventeenth century, Hermann Boerhaave, the leading physician of his time, recommended that an aging patient seek to regain his health by sleeping between two young virgins, recalling King David's futile attempt to do the same kind of thing. History has taken us through the pastoral period of mother's milk and the pseudoscience of monkey glands to rejuvenate flagging juices, and now we are in what might be called the vitamin era, both C and E. Never yet has anyone succeeded in borrowing any time. Most recently, a few researchers have been telling us that growth hormone may hold the promise of increasing lean body mass and bone density, and some among us insist that it will therefore also make people younger. We now hear early rumblings that so-called gene therapy is the answer, whereby cutting and splicing DNA will add decades or more to the maximum life span. In vain do the sober scientists try to convince the clamorers that it just ain't true, and it shouldn't be. The lesson is never learned—there will always be those who persist in seeking the Fountain of Youth, or at least delaying what is irrevocably ordained.

There is a vanity in all of this, and it demeans us. At the very least, it brings us no honor. Far from being irreplaceable, we *should* be replaced. Fantasies of staying the hand of mortality are incompatible with the best interests of our species and the continuity of humankind's progress. More directly, they are incompatible with the best interests of our very own children. Tennyson says it clearly: "Old men must die; or the world would grow moldy, would only breed the past again."

It is through the eyes of youth that everything is constantly being seen anew and rediscovered with the advantage of knowing what has gone before; it is youth that is not mired in the old ways of approaching the challenges of this imperfect world. Each new generation yearns to prove itself—and, in proving itself, to accomplish great things for humanity. Among living creatures, to die and leave the stage is the way of nature—old age is the preparation for departure, the gradual easing out of life that makes its ending more palatable not only for the elderly but for those also to whom they leave the world in trust.

I am not arguing here against an old age that is active and rewarding. I do not advocate going gentle into that enveloping night which is premature senility. Until it becomes impossible, vigorous exercise of body and mind magnifies each living moment and prevents the separation that makes too many of us become older than we are. I speak only of the useless vanity that lies in attempts to fend off the certainties that are necessary ingredients of the human condition. Persistence can only break the hearts of those we love and of ourselves as well, not to mention the purse of society that should be spent for the care of others who have not yet lived their allotted time.

When it is accepted that there are clearly defined limits to life, then life will be seen to have a symmetry as well. There is a framework of living into which all pleasures and accomplishments fit—and pain, too. Those who would live beyond their nature-given span lose their framework, and with it lose a proper sense of relationship to those who are younger, gaining only the resentment of youth for encroaching on its careers and resources. The fact that there is a limited right time to do the rewarding things in our lives is what creates the urgency to do them. Otherwise, we might stagnate in procrastination. The very fact that at our backs, as the poet cautions his coy mistress, we "always hear / Time's wingèd chariot hurrying near" enhances the world and makes the time priceless.

The originator of the literary form we call the essay, the sixteenth-century Frenchman Michel de Montaigne, was a social philosopher who viewed mankind through the scrutinizing lens of unadorned and unforgiving reality and heard its self-deceits with the ear of a skeptic. In his fifty-nine years, he gave much thought to death, and wrote of the necessity to accept each of its various forms as being equally natural: "Your death is a part of the order of the universe, 'tis a part of the life of the world . . . 'tis the condition of your creation." And in the same essay, entitled "To Study Philosophy Is to Learn to Die," he wrote, "Give place to others, as others have given place to you."

Montaigne believed, in that uncertain and violent era, that death is easiest for those who during their lives have given it most thought, as though always to be prepared for its imminence. Only in this way, he wrote, is it possible to die resigned and reconciled, "patiently and tranquilly," having experienced life more fully because of the constant awareness that it may soon come to an end. Out of this philosophy grew his admonition, "The utility of living consists not in the length of days, but in the use of time; a man may have lived long, and yet lived but a little."

V

Alzheimer's Disease

VIRTUALLY EVERY DISEASE can be described in terms of cause and effect. The symptoms a patient presents to his doctor, and the physical findings elicited on examination, are the direct results of very specific pathological changes within cells, tissues, and organs, or of disorders in biochemical processes. Once these underlying alterations have been identified, they can be shown to have led inevitably to the observed clinical manifestations. It is the purpose of the diagnostic workup to find the cause, using its effects as clues.

For example: Atherosclerotic obstruction in the artery that nourishes a segment of heart muscle will cause angina or infarction, with the resultant symptoms of those conditions; a tumor that produces an oversupply of insulin drastically reduces levels of glucose in the blood, preventing proper brain nutrition and leading to coma; a virus that attacks the motor cells in the spinal cord causes paralysis of the muscle to which those cells send messages; a loop of gut becomes twisted around a strand of internal postoperative scar tissue, and the consequent intestinal obstruction produces distension, vomiting, dehydration, and chemical imbalances in the blood, which in turn can lead to cardiac arrhythmia; a ruptured appendix fills the abdominal cavity with pus and the resultant peritonitis floods the bloodstream with bacteria that cause high fevers, sepsis, and shock. The list of examples is endless, and is the stuff of medical textbooks.

The patient comes to the doctor with one or more signs or symptoms—angina, or coma, or paralyzed legs, or persistent vomiting and a swollen belly, or fever accompanied by abdominal pain—

and the detective work begins. It is to the series of events that has led to the observable set of symptoms and other clinical findings that the physician refers when he uses the term *pathophysiology*.

Pathophysiology is the key to disease. To a physician, the word has connotations that convey both the philosophy and the aesthetic of poetry—not surprisingly, part of its Greek root, *physiologia*, has a philosophic and poetic meaning: "an inquiry into the nature of things." When *pathos*—"suffering" or "disease"—is prefixed to it, we have a literal expression of the essence of the doctor's quest, which is to make inquiry into the nature of suffering and disease.

It becomes the doctor's job to identify the instigating cause of sickness by tracing back along the sequence until he has found the ultimate culprit—microbial or hormonal, chemical or mechanical, genetic or environmental, malignant or benign, congenital or newly acquired. The investigation is done by following the clues left in the identifiable damage done to the body by the perpetrator. The crime is thus reconstructed and a treatment plan devised that rids the patient of the influence of the instigator of the disease.

In a sense, then, every doctor is a pathophysiologist, an investigator who identifies the disease by tracing the origins of its symptoms. That having been done, appropriate therapy can be chosen. Whether the aim is to excise the pathology, destroy it with drugs or X ray, counteract it with antidotes, strengthen the organs it is attacking, kill its causative germs, or simply to hold it in check until the body's own defenses can overwhelm it, a plan of action must be organized against each disease if the patient is to stand any chance of overcoming it. When a physician engages in combat to struggle against his patient's mortality, his knowledge of cause and effect is the armory to which he turns to help him choose his weapons.

The result of this past century's biomedical research is that the pathophysiology of the great majority of diseases is well known, or at least known well enough so that effective treatment is available. But there remain some diseases in which the relationship between cause and effect has been less clearly delineated than we might hope, and a few of these diseases are among the greatest scourges of our time. The malady which these days is called "senile dementia of the Alzheimer type" not only falls into this category but carries the additional vexation that its primary cause has continued to elude scientists since the problem was first brought to medical attention in 1907.

The fundamental pathology of Alzheimer's disease is the progressive degeneration and loss of vast numbers of nerve cells in those portions of the brain's cortex that are associated with the so-called higher functions, such as memory, learning, and judgment. The severity and nature of the patient's dementia at any given time are proportional to the number and location of cells that have been affected. The decrease in nerve-cell population is in itself sufficient to explain the memory loss and other cognitive disabilities, but there is another factor that seems to play a role as well—namely, a marked decrease in acetylcholine, the chemical used by these cells to transmit messages.

These are the basic elements of what is known about Alzheimer's disease, but they are far too few to provide a direct linkage between structural and chemical findings on the one hand and the specific manifestations presented at any given moment by the patient on the other. Many of the details of the pathophysiology of the disease still elude the most determined efforts of medical science to pin them down. The sequential features in the long lists of causes, effects, and treatments that appear in the foregoing paragraphs have no analogy to the present state of our knowledge (or ignorance) of Alzheimer's. We know not a whit more about what might cure it than we do about what might cause it.

Consequently, in the course of describing how Alzheimer's disease kills its victims, it will not be possible to stop here and there during the narration of a downhill course to correlate specific symptoms with the stages of pathophysiology of which they are manifestations. Such explanatory digressions would be unsatisfactory and confusing. But there are some very interesting things it *will* be possible to do, and I present them here in yet another list: It *will* be possible to describe the fundamental pathological changes that occur in the brain, and some of the areas of study by which attempts are being made to elucidate them; it *will* be possible to use the gradual historical development of our present knowledge of the illness in such a way that the often abstruse aspects of disordered brain function may be made comprehensible; it *will* be possible to chronicle the emotional carnage visited on the families of victims; it *will* be possible to tell what happens to an afflicted person—and how he or she dies.

"Everything came to a head just ten days before our fiftieth wedding anniversary." Janet Whiting was recalling the six tormented years of her husband's agonizing decline into the final stages of Alzheimer's disease. I have known Janet and her husband, Phil, since my boyhood. They were young and very attractive newlyweds the first time my family visited their apartment in the late 1930s; he was twenty-two and she was twenty. Compared with my immigrant parents, who were staidly ensconced in their forties, the Whitings appeared like a movie-star couple, a pair of juveniles not old enough to be doing anything in that recently furnished apartment but playing house.

Not that I doubted the reality of the excitement Janet and Phil very obviously felt about each other—what I doubted was the likelihood that a couple whose shared life was so joyous could really be married. I was sure they were just trying it out; I knew from personal observation that married people don't behave like this. If the Whitings expected things to work out, they would simply have to stop acting as though they were crazy about each other.

To a great extent, they never did. There remained in that marriage a certain mutuality of gentle regard that I learned increasingly to value as I grew old enough to know something about how it is between a man and a woman. Even the overt unselfconscious expressions of affection never disappeared. As the years passed, Phil made a prosperous career in commercial real estate, and the Bronx apartment was in time succeeded by the beautiful house in Westport, Connecticut, where the three Whiting children were raised. After the kids were grown, Janet and Phil moved to a luxurious condominium in Stratford. When Phil retired from full-time work at sixty-four, the children were long since successfully on their own, there was plenty of money, and the future seemed secure.

After several decades of not seeing the Whitings between my

early twenties and forties, my path crossed theirs again in 1978, when they were living in the condominium, not far from my home near New Haven. To spend an evening with those two great-hearted people was to admire the equanimity of their relationship and the tender respect that was implicit in even their slightest references to each other. Their union had more than fulfilled the promise of its first months. When Phil finally retired completely, and he and Janet made a permanent move to Delray Beach, Florida, my wife and I felt that we had been wrenched away from two valued friends. What we didn't know was that small strange things had already begun to happen.

Even before the move, Phil, whose keen mind had always soaked up works of nonfiction in every spare moment, had stopped reading books. Only in retrospect did that seem strange to Janet, and only in retrospect she found herself years later understanding why he began to insist that she arrange her day so that he might never be out of her company. "I didn't retire," he would grumble when she was leaving to spend an afternoon in town, "to be alone." In his earlier days, outbursts of anger had been rare; now they became more frequent, and turned into full-blown temper tantrums during those last few years in Stratford; increasingly, Phil seemed to find reasons to criticize his daughter Nancy—her visits to the condominium usually ended in tears before she got back on the train to return to her apartment in New York City. After the move to Florida, unexplainable episodes of confusion took place with mounting frequency, and Phil responded to them with disbelief and anger, as though someone else were always at fault. For example, more than once he went to the wrong shop for his regular haircut, then berated the blameless barber for supposedly neglecting the appointment he had really made elsewhere. On one occasion, he threatened to punch a startled motorist at a gas pump, just because the fellow was reaching for an adjacent fuel nozzle—this from a man who had never in his life raised a hand in anger.

Finally, there appeared the first major clue that these new failings were not merely the worsening idiosyncrasies of an aging executive restlessly unfulfilled in his retirement. One evening, Janet invited to dinner a couple whom she and Phil had not seen for several years, Ruth and Henry Warner. Phil had always been an affable host, proud of his wife's table and his own extensive knowledge of wines. Having grown somewhat corpulent while still a young man, he had learned to wear his girth well, so that his ample belly and easy round-faced smile contributed to an air of cheerful prosperity that fairly exuded from some bountiful generosity of spirit within. He was an easy man to like, and he knew how to expand the atmosphere of comfortable bonhomie that his very presence suggested. In his own home or someone else's—it made no difference—Phil was like a bighearted innkeeper whose only wish was the well-being of everyone around him.

And so it had been at that dinner. Janet's food was delicious, Phil's wines were expertly chosen, the table talk was by turns intense and lighthearted, and the evening was enveloped in the cozy mist of *gemütlich* pleasure that was typical of a visit to the Whiting home. The Warners said their good nights in the warm haze of that good feeling so well remembered from earlier years.

On the following morning, Phil couldn't remember any of it. He was unaware of having so much as seen the Warners, and no amount of explanation would convince him that they had visited. "And that frightened me," recalled Janet, whose mind until that hour had been seeking rationalizations for the undeniable changes in Phil's recent behavior. And yet, even at that morning's point of seeming no return, she tried to explain away this most recent of the disquieting episodes she was so often observing. "I thought, Well, sometimes I forget things, too, and maybe he'll talk about it later." So desperate was she to look away from the glaring terror of thoughts that were mounting ever higher in her awareness that she almost convinced herself of the insignificance of her husband's latest lapse.

But a few weeks later, Janet's fragile structure of defenses was overwhelmed by an incontrovertible demonstration that forced itself in sharp focus into her direct line of sight, refusing to be dispersed or even blurred by her exhausted powers of justification. On returning home from a few hours away one afternoon, she found herself confronted by an outraged Phil, angrily accusing her of having gone to visit her lover. Even more upsetting than the accusation itself was the identity of the putative "lover": Phil's cousin Walter, who had been dead for many years. "At that time, I didn't even know what Alzheimer's was—I only knew that I was scared. Something terrible was happening to Phil, and I couldn't ignore or explain it away anymore."

And yet, as though taking action might finally confirm the inevitable, Janet still hesitated to seek medical verification. Perhaps she kept hoping that Phil was simply going through some passing emotional upheaval or that his bouts of inappropriate behavior would not progress, or might even disappear with the passage of healing time. The episodes were, after all, not only brief but unremembered. As soon as the moment passed, Phil seemed unaware of what he had said or done. Thinking back on it even now, Janet doesn't recall the many little lies she must have told herself both to calm the mounting anxiety that was constantly with her and to delay the official pronouncement of hopelessness.

But finally, it became impossible to continue averting her thoughts from the disintegration of Phil's mind. Increasingly often, he would awaken in the middle of the night, shouting at Janet to get out of their bed. "What are you doing here?" he would exclaim. "Since when does a sister sleep with her brother?" Each time, she would patiently do as he demanded and leave him thrashing about in his anger while she lay awake the rest of the night on the living room couch. He would soon fall back into peaceful sleep, then arise in the morning with no recollection of his outburst.

The point came when the moment could no longer be postponed. One day, about two years after the episode with the Warners, Janet used some now-forgotten subterfuge to convince Phil to visit their physician, having finally convinced *herself*. After taking a careful history and doing the physical, the doctor came out of the examining room and pronounced the name of Phil's sickness. By then, Janet had become somewhat familiar with the characteristics of Alzheimer's disease, but even her anticipation of the diagnosis did not lessen the shock or the sense of doomed finality when she heard the words. She and the doctor decided not to tell Phil. It would not have made any difference if they had told him—he was already by then beyond anything but a temporary comprehension of the implications of such a diagnosis, and would never have retained any attempt to describe it. Within minutes of hearing them out, he would have been just as unaware of his mental state as though they had never spoken.

Some months later, in fact, Janet did tell him. As his bouts of irrationality became more frequent and his lapses of memory more prolonged, she was sometimes unable to control her impatience, and always felt an immediate flush of shame when she treated this good man with a hasty outburst of anger or even a sharp word. Once, after a particularly vexing exchange, she snapped at him, "Don't you see what's wrong with you? Don't you know you have Alzheimer's?" In describing that outburst, she told me, "I felt awful the second it was out of my mouth." But her remorse was unnecessary. It was as though she had remarked on the weather. Phil was no more aware of his plight than he had been before she spoke out. As far as he was concerned, nothing untoward had been happening to him—he could not even remember his own forgetfulness. To any casual acquaintance who had a chance encounter with good old Phil Whiting, he seemed just as well as ever, and that is exactly what he thought he was.

Janet did what almost everyone in her anguished situation does. She determined to take care of Phil herself for as long as she could, and she began to look for books to help her understand the state of mind of people with Alzheimer's disease. There were some good ones, but the best of the lot was the aptly named *The 36-Hour Day*. In it she found statements that confirmed what the doctor had told her days earlier, such as: "Typically the disease is slowly but relentlessly progressive," and "Alzheimer's disease usually leads to death in about seven to ten years, but it can progress more quickly (three to four years) or more slowly (as much as fifteen years)." As she wondered whether she was not simply witnessing the ravages of ordinary senility, Janet came across this sentence: "Dementia is not the natural result of aging."

And so Janet soon came to know this was a real disease she would have to confront, and that it carried with it the inexorable certainty of deterioration and death. *The 36-Hour Day* and the other books taught her about the physical and emotional changes she might expect in Phil, as well as giving her important hints about caring not only for him but also for herself during the years of what were certain to be stress and torment. But in the end, she found that "They're just words—it doesn't really penetrate. It's what's in your heart that makes you able to do this." No matter the extent of her reading, and no matter how she tried to prepare herself for the possibility that, as so directly put in *The 36-Hour Day*, "Sometimes people with dementing illnesses . . . may slam things around [or] hit you," she could never have anticipated the series of events that took matters out of her hands one March evening in 1987, after a year of her devoted care. That was the evening "just ten days before our fiftieth wedding anniversary" when "everything came to a head." This is how she described it to me five years later:

> He didn't know who I was—he thought I had broken into the house and that I was stealing Janet's things. Then he began to push me around and throw different objects at me. He broke some of my antiques because he didn't know what they were. Then he said he was going to call Nancy and tell her what was going on. Well, he did call her, and she knew immediately what was happening. She told him, "Just put that woman on," so he pushed the phone at me and said, "Here—my daughter'll talk to you—she'll tell you to get out!" When I took the receiver, Nancy said, "Mother, leave the house right away. I'm calling the police." As I hung up, Phil grabbed the phone and he also called the local precinct.
>
> Foolishly, I stayed in the house, and he began throwing me around. So I called the police, too. Imagine—three police cars showed up and I was so embarrassed! The policemen came in and I tried to tell them what was going on, but Phil said, "She's not my wife. Come with me, and I'll show you a picture of my wife." With that, he took one of the policemen into the bedroom to show him our wedding picture. Of course, when the cop saw the picture, he said, "This bride looks like your wife, standing right here," and Phil insisted, "She's *not* my wife!"
>
> Meanwhile, our neighbor came in and he recognized her. When she saw what was taking place, she spoke to him very gently: "Phil, you know I love you and I wouldn't lie to you. This woman is Janet—turn around and look." So he did just as he was told. He turned around and looked at me as if he was seeing me for the first time. "Janet," he said, "Thank God you're here! Somebody's been in here trying to steal your clothes." And that was it.

One of the officers coaxed Phil into his car. Phil's "They'll think you're arresting me" was fended off with, "Oh no, they'll just think we're taking a friend for a ride." Phil seemed satisfied with such a simple explanation. He was taken to a nearby hospital, where he stayed until a nursing home placement could be made.

Nancy flew down to be with her mother, and they visited the hospital every day. They were at first surprised at the ease with which Phil fell into the ward routine, but soon realized that he didn't actually know where he was. "He would introduce us to the personnel at the ward desk and tell us they were his secretaries and that the hospital was a hotel he was managing." He usually recognized Janet, but had to be told each time that the younger woman was his daughter. In time, he would come to think Janet was his girlfriend, and finally he would have no idea who she was.

Within a week, a good nursing home was found, and Phil was transferred to it. A few days later, Janet spent their fiftieth wedding anniversary there, at the side of a man who sometimes knew why she had come and sometimes didn't. To the true condition of his demented mind, and to the tragedy that had overtaken his family, he was oblivious.

During the next two and a half years, Janet spent most of almost every day with Phil, except for brief periods of respite insisted upon by her children. They could sense her chronic exhaustion and they knew when her ordeal needed to be interrupted. Even her moments of resentment did not escape their awareness, but they understood those, too, and they forgave her more willingly than she forgave herself. No matter the devotion she brought to her ministrations, her lover and best friend had abandoned her to descend into an unknowing slough.

Janet became a volunteer in the physical therapy department, and for a brief time she took part in the activities of a support group for families of Alzheimer's patients. But support groups can only shoulder so much of one's own burden. Within a short time, Janet knew that each victim of dementia inflicts his own unique form of pain on those who love him and that it takes a unique form of response to sustain every distinct afflicted individual. The three children found it impossible to be witness to the destruction of their adored father, and it was a good thing that this was so. They succored the soul of their mother, seeing to it that she was fed the emotional sustenance to carry out the tasks they knew she must undertake.

Joey, the youngest, somehow gathered up the forces necessary to visit his father twice during his long confinement, but he was neither recognized nor remembered. His visits caused him unbearable distress and helped his father not at all. What helped his mother—and this was the real help that was most needed—was the certainty of the support that comes not from groups and not from books but from the sustained devotion of family and those few friends whose loyalty finds its origins in love.

"It's what's in your heart that makes you able to do this." What was in Janet's heart was to do for Phil what only she—no nurse, no doctor, no social worker—could do. Whether he knew her or not—and in time he did not—something almost irretrievable within him must have retained the dim, far-away comprehension that she was security and certainty and predictability in otherwise uncontrollable, meaningless surroundings. "When he saw me coming, he would wave, but he didn't know who I was. He only knew I was someone who came to see him, who sat with him."

At first, the shock of watching Phil's steady deterioration was horrifying anew each day. Somehow, Janet could maintain her equanimity while she was actually with him, though not always: "At times that first year in the home, I'd go to pieces. And then they would take me into a room and talk to me, and talk to me, until I felt a little better able to cope. But every evening I'd go home and have hysterics." Gradually, she became just enough inured to Phil's steady worsening, but she recognized how difficult it might be for others who cared about him. And she also wanted to protect him, wanted him to be remembered as he had been, a man of vibrant goodwill who carried himself not only with dignity but with a sense of style that was uniquely his own. "I wouldn't let our friends visit the home—I didn't want them to see him like that."

In the home, Phil's course was just as the books had foretold: "slowly but relentlessly progressive." At first, he retained some of his gregarious good nature, apparently in the belief that he was host to a homeful of unwell people for whose welfare he was responsible. Fully dressed, he would go from patient to patient, inquiring with proprietary benevolence of each one, "And how are you today? I hope you're feeling well." Sometimes, if Janet or the nurses were distracted for a brief interval, he would push a wheelchair-bound man or woman right out the front entrance of the building, to go for a stroll. He would then have to be apprehended somewhere on the local streets, cheerfully rolling his complaisant and unknowing charge through the turmoil of passing traffic and pedestrians.

During the middle phases of his illness, Phil had developed a marked incongruity between the thoughts he seemed to be trying to express and the words that actually came out. Although this kind of thing may sometimes happen to stroke victims, they are usually aware of their inability to come out with the right words. Phil had no idea of his disability. Janet remembers one particular occasion when, while they were walking together, he suddenly shot out at her, "The trains are running late—do something!" When she replied that she didn't know where the trains were, he retorted in anger, "What's the matter with your eyes, can't you see?" and pointed down at his untied shoelaces. Suddenly, she understood. "He just wanted me to tie his laces, but he expressed it in this way. He knew what he wanted to say, but he couldn't get the right words, and didn't even realize it."

After he had been in the home a while, Phil began to gain weight, a total finally of forty-five extra pounds on his already-generous proportions. But then he stopped eating, forgot, in fact, how to chew. Janet would have to put her finger in his mouth to extract bits of food lest he choke on them. By that time, he no longer remembered his name. Although his ability to chew returned, he never again knew who he was. Until he stopped talking altogether, he would every once in a while look at Janet, for just a brief moment, with the old gentleness. Choosing exactly the words he had used so many countless times during their half century of life together, he would murmur, with all of the familiar softness and devotion of days long gone, "I love you—you're beautiful, and I love you." As soon as the words were uttered, he always crossed back to the other side, the side of oblivion.

Finally, all contact and all control were lost. Phil became totally incontinent but was quite unaware of it. Although fully conscious, he simply had no idea of what had happened. Urine soaking his clothes and smeared sometimes with his feces, he would have to be undressed to clean off the filth that profaned the pittance of humanness still left to him. "And this was a man," said Janet, "who had always been so proud of his appearance, and so dignified. You might even say he was a prude. To see Phil standing there nude while the aides were washing him, with him not having any idea of what was going on . . ." And then, her eyes reflecting the first moist gleam of beginning tears, she said, "It's such a degrading sickness! If there was any way he could have known what was happening to him, he wouldn't have wanted to live. He was too proud to have been able to tolerate it, and I'm glad he never knew. It was more than anyone should have to bear."

And yet, she herself bore it and never questioned that bearing it was what she would do. She saw her children often, and she sat with other wives and husbands of patients whose sorrow she shared. "We'd sit and cry together. When I got a little stronger, I'd try to help them. You get so that you block certain things out—that's what I taught myself to do." She learned that Alzheimer's, though usually a disease of later life, can strike younger people as well. There was a man in his forties in the home. Only his eyes moved.

Toward the end, Phil began to lose weight rapidly. During the last year of his life, the skin seemed to hang from his face; Janet had to buy him new shoes because his feet shrank by two sizes as he became wizened and smaller, and much, much older-looking. This once robustly healthy man, who throughout his adult life had worn well-tailored size 48 suits, fell to a weight of 139 pounds.

And through it all, he never stopped walking. He walked obsessively, constantly, every moment the ward personnel let him. Janet tried to keep up with his rapid pace but would quickly become tired to the point of collapse—and still he continued. Even when he was so weak that he could barely stand, somehow he found the strength to walk back and forth, back and forth, around the confines of the ward. When too exhausted to continue, he would stagger along until Janet and the nurse grabbed his shoulders and eased him down into a chair, too winded and too weak to go farther.

Once seated, the frail body bent sideways because Phil hadn't the strength to hold himself up any longer. The nurses had to tie him in lest he topple to the floor. And even then, his feet never stopped moving. Sitting there, unaware of the world around him, trussed into a chair by a sash around his waist, out of breath from the effects of his ceaseless effort, he would nevertheless keep moving his feet in a pathetic imitation of rapid walking. He was driven to do it, as if pursuing something he had lost forever. Or perhaps that wasn't it at all. Perhaps something inside him knew the fate that awaits those who are in the terminal phases of Alzheimer's disease, and he was running from it.

During his final month of life, Phil had to be tied into bed at night to prevent him from getting up to resume his incessant walking. On the evening of January 29, 1990, in the sixth year of his illness, puffing breathlessly from the effort of one of his fast, forced marches, he stumbled to his chair and fell to the ground, pulseless. When the paramedics arrived a few minutes later, they tried CPR to no avail and sped him to the hospital, which was right next door. The emergency room doctor pronounced him dead of ventricular fibrillation leading to cardiac arrest, then phoned Janet. She had gone home less than ten minutes before Phil began that final walk to his mortality.

> And when he died, I was glad. I know it sounds terrible to say that, but I was happy he was relieved of that degrading sickness. I knew he never suffered, and I knew he had no idea what was happening to him, and I was grateful for that. It was a blessing—it was the only thing that kept me going, all of those months and years. But it was a horrible thing to watch happening to someone I loved so much. You know, when I went to the hospital after Phil died, they asked me if I wanted to see his body. I said no. My friend, who is a devout Catholic, had gone with me, and she couldn't understand my refusal. But I didn't want to remember that face dead. You have to understand—it wasn't for me that I felt that way. It was for him.

And so ended the destruction of Phil Whiting. Even in the midst of his heartbreaking descent into cerebral atrophy, his family was spared the final scene of withering decay that so often plays itself out on the unknowing victim's body. Not uncommonly, late-stage patients already uncommunicative become immobile, their bodies assuming grotesque positions as they stiffen or slump toward death.

But long before the end, the problems of basic hour-to-hour supervision become insurmountable for most families. With behavior unpredictable, the patient's wandering and destructiveness must be prevented or at least dealt with on those occasions when, alertness notwithstanding, caregivers are eluded and damage is wrought. It is for good reason that the authors of *The 36-Hour Day* chose that title. A momentary relaxation of vigilance may result in physical harm to the patient and others, or a conflict with neighbors that forces action long before a family is prepared for it. Energies dissipate, patience wanes, and even the most determined husband or wife soon finds him or herself taxed beyond seeming endurance. Even the routine aspects of nursing care take on a Sisyphean impossibility that defies the best efforts of the most skilled and dedicated of attendants.

It is not a simple thing to find the kind of facility to which one can, with a complete sense of security, entrust someone who has meant so much in one's own life. Although there are many reasons for inadequacy, perhaps the single most important one is a stark statistic: Alzheimer's disease strikes more than 11 percent of the U.S. population over sixty-five. The total number of Americans affected, including those patients below sixty-five, is estimated at around 4 million. The strain on resources will continue and grow. Projections indicate that by the year 2030, the population of Americans who live beyond sixty-five will reach 60 million or more. With direct and indirect cost of dementias of all sorts having already reached an estimated $40 billion annually (and with most of that involving people with Alzheimer's), the magnitude of the problem is ever more staggering. Is it any wonder that a worried family trying to do the best that it can finds itself so often overwhelmed and in need of guidance?

Fortunately, there do exist, albeit still in insufficient numbers, appropriate long-term care facilities in our country, of the sort Janet Whiting was able to find. Some of them even have what are called respite programs, which provide for a short-term admission in order to allow an exhausted caretaker to have a few days or weeks of relief. There exist a few hospice programs, as well. Whatever the degree of a family's reluctance, long-term admission is often the only way by which some measure of tranquillity can be restored.

As more time passes, patients will gradually slide toward complete dependency. Those who do not succumb to such an intercurrent process as stroke or myocardial infarction will very likely lapse into a condition that has been termed, inhumanly and yet very descriptively, the vegetative state. At that point, all higher brain functions have been lost. Even before then, some patients are unable to chew, walk, or even swallow their own secretions. Attempts to feed may result in spells of coughing and choking that are frightening to watch, especially when the observer feels at fault. This is the period when hard decisions are faced by families, having to do with the insertion of feeding tubes and the vigor with which medical measures should be taken to fend off those natural processes that descend like jackals—or perhaps like friends—on debilitated people.

If it is decided not to tube-feed, death by starvation may be a merciful choice for people who are unconscious or otherwise without sensation of the process. Starvation may well seem preferable to the alternatives, the paralysis and the malnutrition that almost inevitably overtake even the most scrupulously fed of intubated terminally ill people. Incontinence, immobility, and low levels of blood protein make it almost impossible to avoid bedsores, which can become ghastly to look at as they deepen to the point of exposing muscle, tendon, and even bone, coated in layers of foul, dying tissue and pus. When that happens, the psychological trauma on the family is mitigated only a little by the knowledge that its victim is unaware of it.

Incontinence, immobility, and the need to catheterize lead to urinary-tract infections. The inability to acknowledge or swallow secretions causes aspiration of mucus and increases the likelihood of pneumonia. Here again, difficult treatment decisions must be made, involving not only individual conscience but religious beliefs, societal norms, and medical ethics. Sometimes, the best course may be not to make those decisions and to let grim nature have its way.

Once embarked upon, that course may sometimes be very rapid. The great majority of people in an Alzheimer's vegetative state will die of some sort of infection, whether it arises in the urinary tract, in the lungs, or in the fetid, bacteria-choked swamp of a bedsore. In the feverish process that ensues, called septicemia, bacteria rush into the bloodstream, rapidly causing shock, cardiac arrhythmias, clotting abnormalities, kidney and liver failure, and death.

All along the way, family members have been experiencing feelings of ambivalence, helplessness, and crisis. They fear what they are seeing, as well as what they have yet to see. No matter how often they are reminded, many people persist in believing they are permitting conscious suffering. And yet, it is always so hard to let go. Such legal instruments as living wills and durable power of attorney may function as so-called advance directives, but all too often they do not exist; a grieving wife or husband, or children already struggling with family problems of their own, are adrift in a sea of conflicting emotions. The difficulty of deciding is compounded by the difficulty of living with what has been decided.

Alzheimer's is one of those cataclysms that seems designed specifically to test the human spirit. The nobility and loyalty of a Janet Whiting are not unique, may even be, to a greater or lesser extent, the norm. By so much is Janet's behavior not exceptional, in fact, that those who provide the professional help come almost to expect that families will rarely question their own roles in the caretaking process. The cost, of course, is considerable. In terms of emotional damage, of neglect of personal goals and responsibilities, of disturbed relationships, and obviously of financial resources, the toll is unbearably high. Few tragedies are more expensive.

It often seems as though the families of Alzheimer's patients are sidetracked from the broad sunlit avenues of ongoing life, remaining trapped for years each in its own excruciating cul-de-sac. The only rescue comes with the death of a person they love. And even then, the memories and the dreadful toll drag on, and from these the release can only be partial. A life that has been well lived and a shared sense of happiness and accomplishment are ever after seen through the smudged glass of its last few years. For the survivors, the concourse of existence has forever become less bright and less direct.

It is probably a universal teaching of all cultures that putting a name to a demon helps to decrease its fearsomeness. I sometimes wonder whether the real, perhaps culturally subconscious, reason that medical pioneers have always sought to identify and classify specific diseases is less to understand than to beard them. Confrontation, somehow, is safer once we have set a label on a thing, as if the very process makes the vicious beast sit still for a while and appear susceptible to taming; it puts under some element of control what has previously been a wildness of unrestrained terror. When we give sickness a name, we civilize it—we make it play the game by our own rules.

Naming a disease is the first step in organizing against it. Not only is it the scientific community that forms the modern-day equivalent of military circles and squares but the community of patients, families, and lay volunteers, as well. Since the middle third of this century, patients and relatives have shared their problems, and sometimes their expenses, with such groups as the National Foundation for Infantile Paralysis, The American Cancer Society, and the American Diabetes Association. People afflicted with these scourges, and those who care about them, need no longer be alone.

In the case of Alzheimer's disease, it is rarely the patient who recognizes the need for company in the journey through travail. But there is probably no disability of our time in which the presence of support groups can help so decisively to ensure the emotional survival of the closest witnesses to the disintegration. In the United States, there are now almost two hundred chapters and more than a thousand support groups under the umbrella of the Alzheimer's Disease and Related Disorders Association (ADRDA), and similar organizations exist in other countries. They function not only to provide help but also as advocates of increased funding for research and clinical improvements. There is strength in numbers, even when the numbers are only one or two understanding people who can soften the anguish by the simple act of listening.

That anguish consists of many parts, and some of them cannot be dealt with unless with a sympathetic and knowing listener. Is it possible that the burden of this disease does not become a source of resentment and sometimes repugnance to everyone it drags along in its loathsome wake? Can anyone maim a great piece of his or her life without seething? Is there a single person who can forbearingly watch as the object of his or her brightest love involutes into incomprehension and decay?

Each family needs help to understand the viciousness of the attack not only on the patient himself but on those who stand with him. Not that help of any sort should be expected to provide release from the torment—it can only make the suffering understandable and offer some respite from the ordeal. The very knowledge that a family's feelings of rage and frustration are universal and unavoidable, the assurance that understanding ears will listen and understanding hearts will share—these are the realizations that can lift away the loneliness and unjustified feelings of guilt and remorse that magnify the deluge of despair visited on each participant by the spiritual subjugation of Alzheimer's.

The road back from isolation starts with the pronouncement of the words that give the alarming symptoms a name. That very act sets in motion the process by which family members can unite their defenses with the millions of others who walk alongside them. The name of this disease did not exist one hundred years ago, although aspects of the process that would be associated with it had been observed and described for centuries as an as-yet-unspecified part of that vast panorama called senility.

"Dementia of the Alzheimer type" is the official title of the disease that is at present being newly diagnosed in several hundred thousand people each year in the United States. It accounts for somewhere between 50 and 60 percent of all dementing illnesses in those over the age of sixty-five and strikes many others in their middle years. The American Psychiatric Association describes its onset as insidious, with a "generally progressive, deteriorating course, for which all other specific causes have been excluded by the history, physical examination and laboratory tests. The dementia involves a multifaceted loss of intellectual abilities, such as memory, judgment, abstract thought, and other higher cortical functions, and changes in personality and behavior."

The dementia itself is defined as: "Loss of intellectual abilities enough to interfere with social or occupational functioning." Behind the deceptively simple words lie centuries of uncertainty and of blurred definitions and categories.

References to what we now call senile dementia, and even legal decisions related to the condition, have appeared for thousands of years in the literary and historical records of Western civilization. Medical authors since antiquity have described it, and physicians gradually came to recognize that younger individuals as well as the old sometimes develop evidences of impaired judgment and memory, and general intellectual deficits of a progressive nature. The word *dementia* did not appear as a medical term, however, until 1801, when it was introduced by Philippe Pinel, who was at that time the senior physician at Le Salpêtrière, a Paris hospital in which several thousand incurable and chronically ill women shared facilities with hundreds of the sick and insane. Pinel has been called the father of the modern treatment of mental diseases, primarily for the accuracy of his descriptions and classifications of psychiatric syndromes, but also for introducing the heretofore-absent element of human kindness into the care of the institutionalized mentally ill, many of whom had previously been kept in chains. He called his novel principle the "moral treatment of insanity."

Pinel systematized his concepts of mental illness in the 1801 book that has become one of the classic volumes in the annals of psychological medicine, *Traité médico-philosophique sur l'aliénation mentale*. In this text, he described a distinct psychiatric syndrome to which he gave the name *démence* (in English, "dementia"), defined as a kind of "incoherence" of the mental faculties. In one brief paragraph entitled "Specific Character of Dementia," Pinel outlined a group of symptoms that will immediately be recognized by anyone who has cared for a patient with the form of it that is today called Alzheimer's:

> Rapid succession or uninterrupted alternation of isolated ideas, and evanescent and unconnected emotions [unconnected either to each other or to external real events]. Continually repeated acts of extravagance: complete forgetfulness of every previous state: diminished sensibility to external impressions: abolition of the faculty of judgement: perpetual activity.

Pinel could have been describing Philip Edward Whiting. The terms *incoherence* and *unconnected* are particularly appropriate, being so consistent with the disabled networks of brain cells, connections, and chemical transmitters of messages that are now recognized to be the fundamental findings of the disease. Pinel was able to distinguish the dementia he introduced from the commonly observed senility of advancing age.

To many clinicians, *incoherence* served as an excellent clinical synonym for *dementia.* In an 1835 publication entitled *A Treatise on Insanity*, James Prichard, a senior physician to England's Bristol Infirmary, pointed out that patients go through a series of stages as the disease progresses, which he called "the several degrees of incoherence." He recognized four such degrees: impaired memory, irrationality and loss of reasoning power, incomprehension, and finally the loss of instinctive and voluntary action. These observations are useful even today in charting the gradual deterioration of individual patients. Indeed, modern authors do write of several stages of illness, and they are very much the same as Prichard's.

Philippe Pinel's student and intellectual heir was a graduate of the thousand-year-old medical school at Montpellier—Jean Etienne Dominique Esquirol. The observations concerning *démence* in Esquirol's 1838 work, *Des Maladies mentales*, have stood the test of time. Once these are studied, little further discussion of dementia's symptoms is required for those who would acquaint themselves with its clinical course as we see it today. Of his patients, Esquirol said:

> [They] have neither desires nor aversions; neither hatred nor tenderness. They entertain the most perfect indifference towards objects that were once most dear. They see their relatives and friends without pleasure, and leave them without regret. They are uneasy in consequence of the privations that are imposed upon them, and rejoice little at the pleasures which are procured for them. What is passing around them, no longer awakens interest; and the events of life are of little account, because they can connect themselves with no remembrances, nor any hope. Indifferent to everything, nothing affects them. . . . Notwithstanding, they are irascible, like all feeble beings, and those whose intellectual faculties are weak or limited. Their anger, however, is only of a moment's duration. . . .
>
> Almost all who have fallen in a state of dementia, have *some sort of ridiculous habit or passion* [Esquirol's italics]. Some are constantly walking about, as if seeking something they do not find. The gait of others is slow and they walk with difficulty. Others still, pass days, months and years, seated in the same

place, drawn up in bed, or extended upon the ground. This one is constantly writing, but his sentiments have no connection or coherency. Words succeed words. . . .

To this disturbance of the soundness of the understanding, are united the following symptoms. The face is pale, the eyes dull, and moistened with tears, the pupils dilated, the look uncertain, and the physiognomy without expression. The body is now emaciated and slender; and now loaded with flesh. . . . When paralysis complicates dementia, the paralytic symptoms are manifested successively. At first, articulation is embarrassed; shortly after, locomotion is executed with difficulty; and the arms are moved with pain. . . . He who is in a state of dementia imagines not, nor indulges in thought. He has few or no ideas. He neither wills nor determines, but yields; the brain being in a weakened state.

Like all the great French medical professors of his day, Esquirol personally performed autopsies on his patients when they died. In that day of microscopes too primitive for accuracy, he had to depend on gross appearances to make his observations. Nonetheless, the findings were striking:

The convolutions of the brain are atrophied, separated from one another, shallowed or flattened, compressed and small, especially in the frontal region. It is not uncommon, that one or two convolutions of the convexity of the brain are depressed, atrophied, and almost destroyed, and the empty space filled with serum.

Esquirol had thus identified an atrophy of the brain that explained the atrophy of the intellect. His observations were confirmed again and again by later investigators. Microscopic analysis, however, would have to await the studies of Alois Alzheimer.

Medical science underwent many profound changes in the seven decades between the contributions of Esquirol and those of Alzheimer, but none was more important than the development of high-resolution microscopes. Their skilled use of the new optical systems enabled scientists at German medical schools to make many of the great discoveries of the second half of the nineteenth century and the first decade of the twentieth. It was in the German tradition of meticulous microscopy that Alois Alzheimer undertook the study of dementia.

Alzheimer had begun his career primarily as a clinician interested in nervous and mental diseases, but he was well trained in laboratory methods. Already an authority on the clinical aspects of senile dementia, and beginning to be known for the clarity of his descriptions of microscopic pathology, he was invited to work at the University of Heidelberg in 1902 by Emil Kraepelin, a pioneer of experimental psychiatry. When Kraepelin was called to the University of Munich in the following year to take charge of a new clinic and research facility, he took the thirty-nine-year-old Alzheimer with him. Alzheimer's skill with the use of newly developed tissue-staining techniques enabled him to identify the changes in cellular architecture that accompany syphilis, Huntington's chorea, arteriosclerosis, and senility. Perhaps the most distinctive characteristic of Alzheimer's work was his ability, based on his own experience with patients, to correlate microscopic postmortem findings with the symptoms exhibited before death by the unfortunate victims of these degenerative processes. Such correlations are the basic elements in tracing the cause and effect of pathophysiology.

In 1907, Alzheimer published a paper entitled "On a Distinctive Disease of the Cerebral Cortex," reporting the case of a woman who had been admitted to the psychiatric hospital in November 1901. This is the first study of a patient in whom the eponymous sickness was recognized as a singular entity to be differentiated from all others. Except that the language is far more stark, we might just as well be reading Esquirol; except that Alzheimer does not specifically outline the boundaries of each of the "four degrees of incoherence," we might just as well be reading Prichard. Alzheimer presented a fifty-one-year-old woman who had gone through the successive symptoms of jealousy, failure of memory, paranoia, loss of reasoning powers, incomprehension, stupor, and, finally, "After 4½ years of the disease, death occurred. At the end, the patient was completely stuporous; she lay in her bed with her legs drawn up under her, and in spite of all precautions she acquired bedsores."

The description of the patient's clinical course was not the reason Alzheimer reported her case. Such patients had been familiar to physicians even before Pinel and Esquirol, although the two French clinicians were the first to separate them into the new category of dementia. In fact, the term *presenile dementia* had been introduced long before Alzheimer, even as early as 1868, in order to differentiate those patients who were still in their middle years when afflicted. Nor did Alzheimer wish to content himself merely with describing yet another demented cortex that was obviously atrophic on simple inspection with the naked eye. His purpose in the 1907 paper was to tell of what he had found when he sectioned this woman's brain, applied special stains to the thinly sliced segments, and examined them under the microscope.

Alzheimer had discovered that many of the cells of the cortex contained one or several hairlike fibrils, which in other cells merged into groups, of increasing density. At what appeared to be a somewhat later stage, the nucleus and indeed the entire cell disintegrated, leaving only a dense bundle of the fibrils in its place. Alzheimer considered the fact that the fibrils absorbed a different dye than did the normal cell to be evidence that the deposition of some pathological product of metabolism was the cause of the lethal outcome. Between one-quarter and one-third of his patient's cortical cells either contained fibrils or had disappeared entirely.

In addition to the destructive processes in the cells, Alzheimer found numerous microscopic clumps or plaques scattered throughout the cortex, which resisted taking up the dye. In later years, these were shown to be composed of degenerated portions of the intercommunicating nerve extensions called axons, clustered around a core of the protein substance beta-amyloid. To this day, the invariable presence of so-called senile plaques and fibrillary tangles are the primary criteria upon which a microscopic diagnosis of Alzheimer's disease is made.

It has been recognized, however, that neither the amyloid plaques nor the neurofibrillary tangles are found exclusively in Alzheimer's. There is a variety of other chronic conditions of the human brain in which one or the other or both can be identified. Even in normal aging, at least a few of both types of structures appear, though hardly in the large quantities characteristic of Alzheimer's. We will know a great deal more about the process of brain aging once the origins of the plaques and tangles of this disease have been discovered.

Alzheimer was wise enough to recognize that "we are apparently confronted with a distinctive disease process." His mentor, Kraepelin, went one step further: Referring to the new entity in the eighth edition of his textbook in 1910, he christened it "Alzheimer's disease." Kraepelin seemed uncertain about the significance of the comparatively young age of Alzheimer's patient, in view of the fact that her history was so similar to those of people who had previously been put into the category of senile dementia. He wrote, "The clinical significance of Alzheimer's disease is still unclear. Although the anatomical findings suggest that this condition deals with an especially severe form of senile dementia, some circumstances speak against this; namely the fact that the disease may arise even at the end of the fifth decade. One would describe such cases at least in terms of senium praecox [precocious senility], if not preferably that this disease is more or less independent of age." This uncertainty by a man whom many considered the high priest of organic psychiatry may have influenced later authors to attach more significance to Kraepelin's use of the term *senium praecox* and to have overlooked his suggestion that "this disease is more or less independent of age." Probably as a consequence of this misinterpretation, the notion that Alzheimer's is a *presenile* dementia became fixed in medical nomenclature for more than half a century.

Within a few years of Alzheimer's paper, other workers reported on similar patients. In each case, the clinical course was not unlike that of Alzheimer's original woman, and the autopsies revealed a diffuse atrophy which, while it involved the entire brain, was particularly evident in the cortex. On microscopic examination, great numbers of the senile plaques and fibrillary tangles could be demonstrated. By 1911, there had been twelve such additional reports.

The relative youth of some of the patients seemed to set the findings off from later descriptions of autopsies in which senile plaques and fibrillary tangles were being found in people of all ages and apparently with a variety of clinical histories. By 1929, there were four reports of the disease in patients below the age of forty, and even one whose symptoms began when he was seven. The problem may have been compounded by a certain selectivity in reporting—physicians are more likely to write up cases that seem unusual than those that are run-of-the-mill. Also, in those countries (and they are in the majority) where autopsies are not mandatory, by far the greater number are done on patients who are "interesting." What is more interesting than a young man with an old man's disease? Thus, by the late 1920s, the great majority of the many cases of Alzheimer's disease in the world medical literature were patients in the relatively young fifty-to-sixty age group.

Although perceptive clinicians obviously appreciated that the age criteria continued to have fuzzy margins, the syndrome continued to be designated as Alzheimer's presenile dementia for decades. That was the name by which I first encountered it in textbooks during my medical school years in the 1950s.

The process by which "Alzheimer's presenile dementia" was transformed into the much more accurate "senile dementia of the Alzheimer type" is a tale paradigmatic of the way biomedical culture has evolved in the last third of the twentieth century. By this, I mean a combination of science, government involvement, and a factor that may best be understood by the term *consumer advocacy*. For sixty years following Alzheimer's initial work, evidence slowly accumulated that there is little or no validity in differentiating between the senile and presenile forms of a disease when both are characterized by the same microscopic pathology. After the point was finally hammered home at a 1970 conference on Alzheimer's and related conditions, there began to arise a scientific consensus that the persistence of such an arbitrary distinction was not only erroneous but misleading.

One of the obvious effects of the changed attitude was the inclusion under the diagnostic umbrella of a perfectly huge population of elderly patients and their families. As research interests were spurred, scientists quite correctly began to clamor for more funding and sought it from government sources. In the United States, this meant the involvement of the National Institutes of Health (NIH) and the enlistment of every advocate of the elderly who might have some political influence. The creation of the National Institute on Aging (NIA) was the natural outgrowth of this process. The coordinating of the efforts of the scientists, the NIA, and caregivers resulted in the founding of the ADRDA. A malady thought in my medical school days to be so unusual that it was used as a trivia question in late-night study sessions had emerged as one of the leading causes of death in World Health Organization statistics. As a result of all the coordinated effort, the Alzheimer's research budget in the United States in 1989 was some eight hundred times what it had been only ten years earlier.

In spite of the great progress that has been made during the last decade and a half in the care of patients and the support of those who must provide that care, the advancements in the more biomedical aspects of the disease, such as cause, treatment, and prevention, have not yet led to the discovery of any distinct cause of the disease, a method of curing it, or any way in which it might be prevented.

There is some evidence that there may be a genetic predisposition to Alzheimer's, but it is less than convincing with regard to older patients and not yet satisfactorily proven for the younger, even though certain chromosomal defects have been identified in small numbers of people with the disease. Explorations into the effect of external factors such as aluminum and other environmental agents, viruses, head trauma, and decreased sensory input sometimes yield suggestive findings and other times do not. As in other maladies of obscure etiology, changes in the immune system have been studied without definitive outcome, and even that ubiquitous villain the cigarette has been suspected by some. What seems very likely is that there will prove to be a range of different pathways each of which leads eventually to the degenerative process of Alzheimer's.

Certain physical and biochemical changes have been found to be accompaniments of the disease process, but their role is still unclarified. For example, biopsy of a patient's cerebral cortex demonstrates a 60 to 70 percent decrease in levels of acetylcholine, a key factor in the chemical transmission of nerve impulses. In fact, attempts to find some effective treatment have focused to a large extent on the search for drugs that might improve the defects in neurotransmission.

Evidence has recently appeared to indicate that acetylcholine may have a role in regulating the body's production of amyloid.

It appears that amyloid increases when levels of acetylcholine are low. This finding provides a possible direct link between the chemical characteristics of the disease and its microscopic pathology, and it may lead to new methods of treatment. Especially provocative has been the suggestion that beta-amyloid is toxic to nerve cells; if this still-controversial idea can be substantiated, there will probably be some real reason for optimism in the search for effective therapy. To illustrate the degree of scientific controversy, it must be appreciated that neurobiologists continue to disagree over the question of whether amyloid causes the degeneration of nerve cells or is merely the result of the breakdown of those cells.

Also, a third microscopic characteristic has been added to the duo of fibrillary tangles and senile plaques, which is the presence within certain cells in the hippocampus of empty spaces called vacuoles, surrounding densely stained granules of uncertain significance. *Hippocampus* is the Greek word for seahorse, bestowed by the physicians of antiquity on this graceful curving structure within the temporal lobe of the brain because its elongated shape evoked the image of that peculiar animal. The hippocampus is involved with the storage of memories. Other of its functions have remained enigmatic, and no one is quite sure of the significance of the vacuoles and their contained granules.

And so the laboratory scientists remain puzzled and hard at work. It is difficult to believe, considering the vast amount of research being done and the many findings already being scrutinized, that the present state of knowledge is not the prelude to a period when the small discoveries will begin to coalesce into some very large ones. That is, after all, the way science usually works in this last third of the twentieth century, rather than by huge leaps forward.

Physicians are now at the point where they can make the diagnosis accurately in about 85 percent of cases without resorting to such extreme measures as biopsies of the brain. Among the several important reasons for early diagnostic efforts is the very direct one that there are certain treatable entities which exhibit enough of the characteristics of dementia that they may be confused with it, thereby compounding the tragedy. Among them are depression, medications, anemia, benign brain tumor, low thyroid function, and some of the reversible effects of trauma, such as blood clots pressing on the brain.

There are no consolations in the diagnosis of Alzheimer's disease. The anguish may be mitigated by good nursing care, support groups, and the closeness of friends and family, but in the end it will be necessary for patient and loved ones together to walk through that very tortuous valley of the shadow, in the course of which everything changes forever. There is no dignity in this kind of death. It is an arbitrary act of nature and an affront to the humanity of its victims. If there is wisdom to be found, it must be in the knowledge that human beings are capable of the kind of love and loyalty that transcends not only the physical debasement but even the spiritual weariness of the years of sorrow.

VI

Murder and Serenity

"MAN IS AN obligate aerobe": There, stated with the simple directness of any of the most quoted aphorisms of ancient Hippocrates, stands the secret of human life. The dependence on air of all mankind, and indeed all known terrestrial animals, was recognized by primitive tribesmen long before any of them were distinguished from their fellows by being called healers. No matter the technological sophistication of ultramodern molecular research, and no matter the increasingly abstruse terminology of its current literature, the circle of knowledge always returns to its starting point: In order to live, man must have air.

In the late eighteenth century, it was found that not air in general but one particular component of it, oxygen, is the crucial factor on which life depends. The conception of man as an obligate aerobe then took on a more specific meaning: We have no choice—without oxygen, our cells die and we die with them. Oxygen absorption was soon thereafter shown to be the reason that the color of blood turns instantly from a dark tiredness to the bright red of vibrant life as it passes through the lungs; its departure into the cells of the body's distant tissues was recognized as the cause of blood's exhaustion when it returns depleted and blue from the long journey, figuratively gasping for air. Since then, the role of this most vital of nature's elements has been explored generation after generation by thousands upon thousands of researchers, who have recorded their findings in virtually every one of the world's written languages. Oxygen is at the focal point of the lens through which the sustaining processes of living things must be studied.

After all the years and all the research, the scholars of human biology come ever back to the few words that have always been inherent in an individual's understanding of what he must do to stay alive: Man is an obligate aerobe. I could have plucked one of the many variations of that maxim from almost any of the past two centuries' profusion of writings on the subject, but its actual source is instructive. I found it in a recent issue of the *Bulletin of the American College of Surgeons*, entitled "What's New in Surgery—1992." It appeared not as the time-honored nutshell of wisdom it is but as an experimentally proven, molecular-level certainty. What may be even more revealing are the statement's surroundings; it is situated smack in the middle of the *Bulletin*'s highly technical article on the latest developments in critical care, that brand-new (the trendy term is *cutting edge*) superspecialty created to defend the very border of a desperately ill person's flickering existence, the ultimate battleground contested between the strained forces of life and the powerful assaults that disease is launching in order to overwhelm them.

The new specialty's venue is the intensive care unit; its primary defensive strategy is to maintain a dependable supply of oxygen to the beleaguered cells of the body. Certainly our cave-dwelling forebears would have agreed that this is the right thing to do. The late Milton Helpern, to whose autopsy rooms patients were sent for study if the battle was lost, spent his career seeking out the "ten thousand several doors" to death, and he always came up with the same underlying answer: not enough oxygen.

Oxygen takes a remarkably direct route in making its way from the inhaled air to its ultimate destination, which is the aerobically obligated cell. After passing readily through the thin walls of the lung's alveoli and their attached network of capillaries, the oxygen molecules link themselves to the protein pigment of the red cells which we call hemoglobin. Thereafter known as oxyhemoglobin, the combined molecules are carried from the lung to the left heart and then out through the aorta to the broad highways and narrow footpaths of the arterial circulation, until they reach the distant capillaries in the tissues whose sustenance is the object of their journey.

Once arrived, the oxygen separates itself from its traveling companion, hemoglobin. It leaves the red cell like a passenger getting off a railroad car, and enters the individual tissue cell along with biochemical substances required for that cell's normal function. In what may be thought of as an exchange, carbon dioxide diffuses into the circulating blood, which also carries away the waste products of cellular life, to be destroyed or released through those magnificently multitalented organs of purification, the liver, the kidneys, and the lungs.

Like any good system of delivery and pickup, this one depends on a predictably consistent flow of traffic, in this case the traffic being blood. *Shock* is the term used to describe the course of events that ensues when the blood flow is inadequate to meet the needs of the tissues. Although shock may be caused by a variety of mechanisms, the majority of cases are due to failure of the heart's pumping action (as in myocardial infarction) or to a major decrease in circulating blood volume (as in hemorrhage). The two mechanisms are called, respectively, cardiogenic and hypovolemic shock. Another common instigator of shock is septicemia, the entrance into the bloodstream of the products of infection. So-called septic shock has profound effects on cellular function, as will be discussed later, but one of its major actions is to induce a redistribution of blood so that it pools in certain extensive networks of veins, like those of the intestine, thereby becoming lost to the general circulation. Regardless of cause, all forms of shock have a similar outcome: Cells are deprived of their source of biochemical exchange and oxygen, the ultimate factor in their death.

Whether or not they do die, and whether enough of them die so that the patient is also killed, is determined by the duration of the shock. If it lasts long enough, it is always lethal. The *long enough* is, of course, a relative term. Just how long is long enough? It depends upon the degree of the circulation's inadequacy. If flow is stopped completely, as in cardiac arrest, death occurs within minutes; if it is merely decreased to levels somewhat less than those needed for survival, dying takes more time and occurs at different rates in different tissues, depending upon how much oxygen their cells require. The brain being particularly sensitive to deficiencies of oxygen and glucose, it fails quickly; because its viability is the legal criterion of being alive, there is obviously a very narrow margin between mortality and continued existence in those people whose cerebral circulation is at all compromised. Interference with oxygen delivery to the brain is a factor in a wide variety of violent deaths.

Although viability of the brain is currently the legal criterion by which mortality is determined, there is still usefulness in the time-honored way in which clinical physicians have always diagnosed death. *Clinical death* is the term used to encompass that short interval after the heart has finally stopped, during which there is no circulation, no breathing, and no evidence of brain function, but when rescue is still possible. If this stoppage occurs suddenly, as in cardiac arrest or massive hemorrhage, a brief time remains before vital cells lose their viability, during which measures such as cardiopulmonary resuscitation (CPR) or rapid transfusion may succeed in resuscitating a person whose life has seemingly ended—the time is probably no more than four minutes. These are the dramatic moments we read about and see portrayed on our television screens. Although the attempts are usually futile, they succeed just often enough that, under the appropriate circumstances, they should be encouraged. Because individuals most likely to survive clinical death are those whose organs are healthiest and who do not have terminal cancer, for example, or debilitating arteriosclerosis or dementia, their continuing existence is still possible and potentially most valuable to society, at least in terms of ability to contribute. It is for this reason that the principles of CPR should be taught to every motivated person.

Clinical death is often preceded (or its first evidences are accompanied) by a barely more than momentary period termed the *agonal phase*. The adjective *agonal* is used by clinicians to describe the visible events that take place when life is in the act of extricating itself from protoplasm too compromised to sustain it any longer. Like its etymological twin, *agony*, the word derives from the Greek *agon*, denoting a struggle. We speak of "death agonies," even though the dying person is too far gone to be aware of them, and even though much of what occurs is due simply to muscle spasm induced by the blood's terminal acidity. Agonal moments and the entire sequence of events of which they are a part can occur in all the forms of death, whether sudden or following upon a long period of decline into terminal illness, as in cancer.

The apparent struggles of the agonal moments are like some violent outburst of protest arising deep in the primitive unconscious, raging against the too-hasty departure of the spirit; no matter its preparation by even months of antecedent illness, the body often seems reluctant to agree to the divorce. In the ultimate agonal moments, the rapid onset of final oblivion is accompanied either by the cessation of breathing or by a short series of great heaving gasps; on rare occasions, there may be other movements as well, such as the violent tightening of James McCarty's laryngeal muscles into a terrifying bark. Simultaneously, the chest or shoulders will sometimes heave once or twice and there may be a brief agonal convulsion. The agonal phase merges into clinical death, and thence into the permanence of mortality.

The appearance of a newly lifeless face cannot be mistaken for unconsciousness. Within a minute after the heart stops beating, the face begins to take on the unmistakable gray-white pallor of death; in an uncanny way, the features very soon appear corpselike, even to those who have never before seen a dead body. A man's corpse looks as though his essence has left him, and it has. He is flat and toneless, no longer inflated by the vital spirit the Greeks called *pneuma.* The vibrant fullness is gone; he is "stripped for the last voyage." The body of the dead man has already begun the process of shrinking—in hours, he will seem "to be almost half himself." Irv Lipsiner reenacted the deflation by blowing his breath out through pursed lips. No wonder we say of the recently deceased that they have expired.

Clinical death has a distinctive look about it. A few seconds' observation of the victim of cardiac arrest or uncontrolled hemorrhage will decide the appropriateness of attempts at resuscitation. Should any doubt remain, there are the eyes to consider. If open, they are at first glassy and unseeing, but if resuscitation does not commence they will in four or five minutes yield up their sheen and become dulled, as the pupils dilate and forever lose their watchful light. It is soon as though a thin cloud-gray film has been laid down over each eye, so that no one can look within to see that the soul has fled. Its rounded plumpness having depended on something no longer there, the eyeball soon flattens out, just enough to be noticeable. It is a flatness from which there is no rising.

The absence of circulation is confirmed by the absence of pulse—an observer's seeking finger on the neck or groin detects no sign of a throbbing artery beneath, and the surrounding muscles, if they are not still in an element of spasm, have begun to assume the flaccid consistency of meat slabbed in a butcher's display case. The skin has lost its elasticity, and that slight shine is gone which once gleamed in reflected recognition of nature's light. At that point, life is over—no amount of CPR can retrieve it.

To be declared legally dead, there must be incontrovertible evidence that the brain has permanently ceased to function. The criteria of brain death currently being used in intensive care and trauma units are very specific. They include such signs as loss of all reflexes, lack of response to vigorous external stimuli, and absence of electrical activity as shown by a flat electroencephalogram for a sufficient number of hours. When these standards have been met (as when brain death is due to head injury or massive stroke), all artificial supports can be withdrawn and the heart, if not already stilled, will soon stop, ending all circulation.

When circulation ceases, cellular death can complete itself. The central nervous system goes first and the connective tissue of muscle and fibrous structures goes last. With electrical stimulation, it is sometimes possible to induce muscular contraction even hours after death. Some few organic processes, called anaerobic because they require no oxygen, will also continue for hours, such as the liver cell's ability to break down alcohol into its component parts. The supposedly well-known fact that hair and nails will keep growing for varying periods of time after death is not a fact at all—no such thing happens.

In most deaths, the heartbeat ends before the brain ceases to function. Particularly in sudden deaths due to trauma other than head injury, the cessation of heartbeat is almost always the result of the rapid loss of more blood than can be tolerated—the trauma surgeon refers to such a hemorrhage as *exsanguination*, which is a more elegant term than the more commonly used *bleeding out.* Exsanguination may be due to direct laceration of a major vessel or to injuries of blood-filled organs like the spleen, liver, or lung; sometimes the heart itself is torn.

The rapid loss of approximately one-half to two-thirds of the body's blood volume is usually sufficient to arrest the heart. Since total blood volume is equal to some 7–8 percent of body weight, a bleed of eight pints in a 170-pound man or six pints in a 130-pound woman can be enough to cause clinical death. With laceration of a vessel the size of the aorta, the process takes less than a minute; a tear in the spleen or liver might take hours, or even days, on those very rare occasions when constant ooze remains unchecked.

With the loss of the first few pints, blood pressure begins to drop and the heart speeds up in an attempt to compensate for the decreased volume of each stroke. Finally, no amount of internal readjustment can keep up with the losses—the pressure and volume of blood reaching the brain become too low to sustain consciousness, and the patient lapses into coma. The cerebral cortex fails first, but the brain's "lower" parts, such as the brain stem and medulla, hold on a bit longer, so that respiration continues, though in an increasingly disorganized fashion. Finally, the near-empty heart stops, sometimes fibrillating before it does so. The agonal period then begins, and life flickers out.

This entire grim sequence of events—hemorrhage, exsanguination, cardiac arrest, the agonal moments, clinical death, and finally irretrievable mortality—was played out during a particularly vicious murder committed a few years ago in a small Connecticut city not far from the hospital where I work. The attack took place at a crowded street fair, in full view of scores of people who fled the scene in fear of the killer's maniacal rage. He had never laid eyes on his victim before the instant of the savage onslaught. She was a buoyant, beautiful child of nine.

Katie Mason was visiting the fair from a nearby town, along with her mother, Joan, and her six-year-old sister, Christine. Accompanying them was Joan's friend Susan Ricci, who had brought along her own two children, Laura and Timmy, about the same ages as the Mason kids—Katie and Laura, in fact, were fast friends and had been studying ballet together since they were both three. As they milled around with the crowd at a sidewalk sale in front of the local Woolworth's, little Christine began tugging at her mother's hand to attract her attention to the pony rides on the other side of the street, begging to be taken over there. Leaving Katie with the others, Joan and her younger daughter crossed the road toward the concession. Just as they reached the opposite sidewalk, Joan heard a hubbub from somewhere behind her and then a child's shrill scream. She turned, dropped Christine's hand, and advanced a few feet toward the sound. People were scattering in all directions, trying to get away from a large, disheveled man who stood over a fallen little girl, his outstretched right arm pummeling furiously away at her. Even through the haze of her frozen incomprehension, Joan knew instantly that the child lying on her side at the crazed man's feet was Katie. At first, she saw only the arm, then realized all at once that in its hand was clutched a long bloody object. It was a hunting knife, about seven inches long.

Using all his strength, up and down, up and down, in rapid pistonlike motions, the assailant was hacking away at Katie's face and neck. In an instant, everyone had fled—murderer and victim were suddenly alone. Unhindered at his frenzied work, the man first crouched and then sat alongside the child, chopping with those ceaseless plungings of his ferocious arm. As the pavement reddened with her child's blood, Joan, by then also alone, stood about twenty feet away, rooted there by disbelief and horror. She would later remember that the air seemed too thick to let her move through it—her body felt warm and benumbed, and she was enveloped in a dreamy mist of insulation.

Except for the ferocious chopping of that unremitting arm coming down again and again on the silent child, there was almost no movement in the entire unearthly scene. Anyone watching from inside the Woolworth's or the refuge of some other concealment might have seen a grotesque tableau of madness and slaughter being enacted on that soundless street.

Though Joan was certain the macabre scene would have no end, her fixed immobility could not have lasted more than a few seconds, but during that seeming protraction of time she saw the knife repeatedly enter her child's face and upper body. Two men suddenly appeared from somewhere beyond the margins of the tableau and grabbed at the killer, shouting as they tried to wrestle him down. But he could not be deterred—with psychotic determination, he kept stabbing at Katie. Even when one of the men began aiming powerful heavy-booted kicks at his face, he seemed not to notice, though his head was being knocked from side to side by the force of the blows. A policeman ran up and seized the knife-wielding arm; only then did the three men manage to subdue the struggling maniac and pin him to the ground.

As the crazed attacker was pulled off Katie, Joan rushed forward and took her daughter into her arms. Turning her gently over from her side onto her back, and looking into that lacerated little face, she said softly, "Katie, Katie" as if she were cooing to a cradled babe. The child's head and her neck were covered with blood and her dress was soaked in it, but her eyes were clear.

> She was gazing at me and beyond me, and there was a warm feeling in me. Her head had fallen back. Then I raised her a bit, and I thought she was still breathing. I spoke her name a few times and told her I loved her. And then I knew that I had to take her to a safe place—I had to get her away from this man, but it was already too late. I lifted her up in my arms. I carried her that way a short distance, and then I thought, What am I doing? Where am I taking her? I got on my knees and very gently put her down. Her chest began heaving and she started to vomit blood. It came out in such huge amounts, constantly—I didn't think she would have so much blood in her; I knew she was emptying out the blood in her body. I screamed for help, but there was nothing I could do to stop the vomiting.
>
> When I had first gone to her, I saw some glimmer in her eyes, almost like some sort of recognition. But by the time I laid her on the ground, her eyes had a different look. Even when she was vomiting blood, they had changed to a more glassy look. When I first went to her side, she still looked alive—but not anymore.
>
> There was no look of pain in her eyes, but instead it was a look of surprise. And then when things changed, she still had that expression on her face, but her eyes had glazed over a little bit. A woman came over—I guess she was a nurse. She started CPR. I didn't say anything, but I thought to myself, Why is she doing that? Katie is not in her body anymore. She's behind me, up there above me, and floating. Her life isn't inside her anymore, and she's not coming back. Her body is just a shell now. At that point, everything was different than it had been when I first went to her side—I had an awareness that my daughter had died. I felt she was no longer in her body, that she was somewhere else.
>
> The ambulance came, and they lifted her out of the pool of

blood and tried to force air into her lungs with an Ambu bag. Her eyes were still wide open and she still had that glassy look. The look on her face was a look of utter surprise, like "What's happening?" It was a combination of being helpless, confused, and surprised, but definitely not a look of horror, and I remember being relieved that it wasn't, because I was looking for any sense of relief at that time. . . .

Later, I went through months and months of asking myself, How much pain did she feel? I needed to know that. I saw her bleed all the blood out of her body when she vomited. Her chest and face were covered with cuts and gashes. She must have been moving her head from side to side, struggling to get free of this man. Later, I found out that he had appeared from nowhere and pushed Laura aside. He had grabbed Katie's hair and thrown her to the ground. It was Laura who screamed, not Katie. I had to know what she went through, what she felt. . . .

Do you know what it looked like? It looked like a release. After seeing him attacking her that way, it gave me a sense of peace to see that look of release. She must have released herself from this pain, because her face didn't show it. I thought, Maybe she went into a state of shock. She looked surprised but not terrified—as terrifying as it was for me, it wasn't that way for her. My friend Susan saw the look, too, and said that maybe Katie had given up, but when I told her I thought it was a look of release, she said, "That's it, you're right!"

We once had a portrait of her made, and it's that same look that she has in her eyes. They were wide but not in a state of terror—it looks almost like an innocence—an innocent release. As her mother, amidst all of that blood and everything else, it was actually soothing to look into her eyes. There came a point when I was with her that I felt like she was out of her body, floating up there looking down on herself. Even though she was unconscious, I felt that somehow she knew I was there, that her mother was there when she was dying. I brought her into the world and I was there when she was leaving—in spite of the terror and horror of it, I was there.

The ambulance sped Katie to the nearest hospital, which was only a few minutes away. Although she was clearly pulseless and brain-dead on arrival, and beyond clinical death, the appalled emergency room team did every possible thing to bring her back, even with the certain foreknowledge that their attempts would be futile. When they finally gave up, their frustration and rage turned to grief. Tearfully, one of the doctors told Joan what she already knew.

The man who murdered Katie Mason was a thirty-nine-year-old paranoid schizophrenic named Peter Carlquist. Two years earlier, he had been acquitted by reason of insanity in the attempted knife murder of his roommate, whom he accused of putting poison gas into their radiator. He had a long history of such attacks on people, including his sister and several high school classmates. As early as age six, he had told a psychiatrist that the devil had come up out of the ground and entered his body. Perhaps he was right.

Following the assault on his roommate, Carlquist had been institutionalized in a unit for the criminally insane on the sprawling grounds of the state mental hospital situated at the outskirts of the city visited by Katie Mason on that fateful July day. Only a short time before, an advisory board had judged him sufficiently recovered to be transferred to a unit housing an assortment of the mentally ill, where patients were permitted to sign themselves out for several hours at a time. On the morning of the assault, Carlquist strolled off the grounds, took a municipal bus into town, and walked into a local hardware store. After buying a hunting knife, he came upon the street fair. And there in the crowd outside Woolworth's he saw the two pretty little girls wearing identical dresses. Somewhere in his deranged mind lies the secret of why he chose the dark-haired Katie to be his victim instead of blond Laura. Rushing forward, he grabbed her by the arm, threw her to the ground, and began his demonic work.

Katie Mason died of acute hemorrhage leading to hypovolemic shock. Although she had been cut in many places on the upper part of her body, the main source of bleeding was a completely severed carotid artery emptying itself into a laceration of her esophagus. The blood passed down the esophagus into her stomach; it was the source of the huge regurgitation.

A specific sequence of events takes place in people who bleed to death. At first, they will usually hyperventilate, which is the body's compensatory attempt to saturate the decreasing volume of circulating blood with as much oxygen as possible; the heart rate will speed up for the same reason. As more blood volume is lost, the pressure in the vessels begins to fall rapidly and the coronary arteries receive less and less of it. Were an electrocardiogram to be running, it would show evidence of myocardial ischemia; the ischemia causes slowing of the poorly oxygenated heart. When the blood pressure and pulse rate become low enough, the brain ceases to receive enough oxygen and glucose, and unconsciousness ensues, preceding brain death. Finally, the ischemic heart slows to a stop, usually without any fibrillation. With the stilling of the heartbeat, circulation is arrested, breathing ceases, there are a few agonal events, and clinical death has occurred. When a vessel the size of the carotid artery has been cut wide open, the entire sequence can take less than a minute.

All of this explains how Katie Mason died. But it does not explain a phenomenon witnessed by her mother, one that matches the descriptions of many other witnesses to such horrendous events. Why should a child who has suddenly been set upon by a knife-wielding psychopath obviously intent on her murder die not only without a look of terror on her face, but actually in a state of apparent tranquillity and release, an appearance of surprise rather than horror? Especially considering the savageries being perpetrated on her face and upper body during the brief time when she must have been fully conscious and seen what was being done to her—why was there no evidence of panic, or even fear?

What was described by Joan Mason has been a source of wonderment, in fact, for hundreds of years. For some soldiers, the absence of pain and fear has been the determining factor in their ability to fight on in spite of crippling wounds, feeling nothing except perhaps the euphoria of battle until the immediate danger is over, whereupon physical and mental agonies finally make their appearance, or death. There is far more at work here than the well-known "fight or flight" of a rush of adrenaline.

In his essay "Use Makes Perfect," Michel de Montaigne suggests that a lifelong acquaintance with the ways of death will soften one's final hours:

> I fancy there is a certain way of making it familiar to us, and in some sort of making trial what it is. We may gain experience, if not entire and perfect, yet such, at least, as shall not be totally useless to us, and that may render us more confident and more assured. If we cannot overtake it, we may approach it and view it, and if we do not advance so far as the fort, we may at least discover and make ourselves acquainted with the avenues.

Montaigne recounts an experience of being thrown from his mount by a horseman "thundering full speed in the very track where I was rushing." Battered and bleeding, he thought at first that he had been shot in the head with a harquebus. But to his surprise, he remained quite calm: "Not only did I make some little answer to questions which they asked me, but they moreover tell me that I was sufficiently collected to order them to bring a horse to my wife, whom I saw struggling and tiring herself on the road."

He describes a sense of tranquillity, even though he refused the soporifics offered to him, "certainly believing that I was mortally wounded in the head. My condition was, in truth, very easy and quiet, I had no affliction on me, either for others or myself; it was an extreme languor and weakness, without any manner of pain." He passed a serene two or three hours awaiting the death that never came, quite content to "glide away so sweetly and after so soft and easy a manner." At the end of that time, "I felt myself on a sudden involved in terrible pain, having my limbs battered and ground with my fall, and was so ill for two or three nights after, that I thought I was once more dying again, but a more painful death."

Whatever was the influence that had so tranquilized the mind of the grievously wounded Montaigne, it had worn off. After that initial period of a few hours, he suffered intense pain. Gone was the serenity, the languor, and the acceptance of an anticipated easy death. The reality of his suffering and fear became inescapable.

Stories like Montaigne's are not rare—they are sometimes given a mystical quality by those who tell them, as though some unexplainable and perhaps supernatural event has taken place. But to a doctor who has spent his career in the company of trauma inflicted in the name of surgical cure, and those other traumas inflicted by the violence of modern life, there is a prototype for these tales of serenity and languorous comfort in the face of what would seem to be frightful and agonizing wounds. The prototype is the aftermath of the injection of an opiate or some other form of powerful narcotic painkiller. When the medication is well chosen and the dose is high enough, fear passes and the distress of even the most unbearable incision or injury recedes into a soft cloud of indifference. Many patients report a sense of well-being, and I have even seen mild euphoria following a proper dose of a morphinelike narcotic.

It is not farfetched to believe that the human body itself knows how to make those morphinelike substances and knows how to time their release to correspond with the instant of need. The "instant of need," in fact, may be the very stimulus that sets off the process.

Such self-generated opiates do, in fact, exist, and they are called endorphins. They were given that name shortly after their discovery about twenty years ago—by contracting the two words that describe them: They are *endogenous morphine*-like compounds. *Endogenous* appeared in the lexicon of medicine at least a century earlier, adapted from the Greek *endon*, meaning "within" or "inner," and *gennao*, meaning "I produce." Accordingly, it refers to substances or conditions we create within our own bodies. *Morphine*, of course, recalls Morpheus, the Roman god of sleep and dreams.

Several structures in the brain are capable of secreting endorphins in response to stress, including the hypothalamus and an area called the periaqueductal gray matter, as well as the pituitary gland. Together with ACTH, a hormone that activates the adrenal glands, endorphin molecules are known to bind themselves, as do the other narcotics, onto foci, called receptors, on the surfaces of certain nerve cells. The effect is to alter normal sensory awareness. Endorphins seem to play a significant role not only in raising pain threshold but also in altering emotional responses. In addition, there is evidence that they interact with the adrenaline-like hormones, as well.

In the normal nonstressed, noninjured person, there is no evidence of the pain-relieving and mood-altering action of endorphins. It requires some definitive degree of trauma, whether physical or emotional, for them to swing into action. The level, or even the quality, of the necessary trauma has not yet been ascertained.

For example, it may be that the mere stimulation of acupuncture needles results in an outpouring of endorphins. During the course of a series of professional travels to Chinese medical schools over a period of years, I became interested in acupuncture after seeing several demonstrations of its effectiveness as an alternative to anesthesia in major surgery. In 1990, I visited Professor Cao Xiaoding, a neurobiologist who heads Shanghai Medical University's Acupuncture Anesthesia and Analgesia Research Coordinating Group, an establishment of thirty faculty members and six laboratories of neuropharmacology, neurophysiology, neuromorphology, neurobiochemistry, clinical psychology, and computer science. Professor Cao's team has produced a vast body of rather impressive experimental and clinical evidence indicating that the basis of acupunture's undoubted success in certain applications is the stimulation of endorphin secretion by manipulation of the vibrating or rotating needles. Although a significant rise in endorphin levels during acupuncture has been repeatedly documented not only in Shanghai but also in several Western laboratories, the neurological pathway by which the turn-on signal reaches the brain has not yet been elucidated. It may be similar to the mechanism that activates the familiar stress-induced response.

Since the late 1970s, endorphins have been shown to make their appearance in the presence of shock due to major blood loss or septicemia; their elevation in physical trauma of all sorts is well documented in the surgical literature. Until fairly recently, this phenomenon had not been studied in children, but a recent report from the University of Pittsburgh demonstrates the same pattern as in adults—namely, a significantly higher increase in endorphins among patients whose injuries were most severe, as compared with those sustaining minor trauma. Some children whose only injuries consisted of abrasions were also shown to have somewhat elevated their levels.

We will never know the level of Katie Mason's endorphins (and some of my proof-demanding clinical colleagues will no doubt find fault with my assumption that it was high), but I am convinced that nature stepped in, as it so often does, and provided exactly the right spoonful of medicine to give a measure of tranquillity to a dying child. Endorphin elevation appears to be an innate physiological mechanism to protect mammals and perhaps other animals against the emotional and physical dangers of terror and pain. It is a survival device, and because it has evolutionary value it probably appeared during the savage period of our prehistory when sudden life-threatening events occurred with frequency. Many a life has no doubt been saved by the absence of panicky response to sudden danger.

Joan Mason, too, seems to have been protected by her endorphins. She told me that had it not been for her own feeling of almost supernal warmth and the sense of being surrounded by a thick insulating aura, she believes that she might have had a heart attack and died there on the street alongside her daughter. The primitive prehuman whose heart and circulatory system did not succumb to sheer terror at the moment of an animal attack was the one who survived to have offspring whose responsiveness was much like his own.

Although there are many narratives of this kind of thing, there has been very little attempt to study it in any systematic way. We read the philosophical lesson of a Montaigne, or a soldier's story, or perhaps the account of a mountain climber who experienced an unaccustomed inner peace while free-falling to an expected sudden death. Some of us have our own tales to tell. And then, of course, there are the times when endorphins fail and death comes in its full unrelieved anguish.

Because to some, endorphins would seem to involve matters of the body, and to others matters of the spirit, it is instructive to examine the experience of an articulate man whose goal was to heal both. Many tend to forget that the great explorer David Livingstone was a medical missionary. He survived a number of close calls during his African forays, but there is one that exemplifies the way in which protoplasm and ectoplasm sometimes work most closely with each other just at the moment when they seem about to part ways forever.

In February 1844, when Livingstone was thirty years old, he was one day set upon by a wounded lion from which he was trying to protect several native tribesmen in his party. The jaws of the enraged animal seized him by the left upper extremity, and he felt himself lifted off the ground and shaken violently as the lion's teeth sank deeply into his flesh, splintering the underlying humerus and ripping eleven jagged lacerations into the bleeding skin and muscle. One of Livingstone's party, an elderly convert named Mebalwe, had the presence of mind to pick up a rifle and discharge both barrels, which sufficiently frightened the animal that he dropped his prey and dashed off, only to die a short distance away of the bullet wound Livingstone had inflicted just before being pounced upon.

The injured explorer had plenty of time to think about his narrow escape during the more than two months it took him to recover from the blood loss, the shattered compound fracture, and the serious infection that began draining pus within a short time. As much as he was amazed at his survival, so also was he by the equanimity he had felt while in the lion's grasp. He would later describe the event and his ineffable sense of peace in the autobiographical work he published in 1857, *Missionary Travels and Researches in South Africa.*

> Growling horribly close to my ear, he shook me as a terrier dog does a rat. The shock produced a stupor similar to that which seems to be felt by a mouse after the first shake of the cat. It caused a sort of dreaminess, in which there was no sense of pain nor feeling of terror, though quite conscious of all that was happening. It was like what patients partially under the influence of chloroform describe who see all the operation but feel not the knife. This singular condition was not the result of any mental process. The shake annihilated fear, and allowed no sense of horror in looking round at the beast. This peculiar state is probably produced in all animals killed by the carnivora; and if so, is a merciful provision by our benevolent Creator for lessening the pain of death.

In that long-ago day when laboratory science was barely beginning its long partnership with bedside medicine, Livingstone's explanation for his remarkable calm was one with which most people probably agreed. It would have taken prescience, or perhaps a disavowal of faith, to have invoked physiology in those dawning moments when microscopy and chemical analysis were but swaddled newborns. For Livingstone to have somehow intuited the principles of stress-related biochemical alteration of states of consciousness was quite improbable. Absent a supreme leap of prophetic vision, beyond the capability of even an ordained Christian missionary, he could not have foreseen the discovery of such a phenomenon.

I have had the personal experience of one such episode. I am not by nature a fearful person, and yet there are two situations that scare me to the point of pathological irrationality: finding myself looking down from some great height, and being immersed in deep water. I need only to think about either of those two hazards to set off a spasm of tightness in each of my sphincters, from the top of the alimentary tube right down to its very end. It is not just that I am cautious about deep water, or even afraid of it—I am unmanned by it, reduced to craven, phobic cowardice. In a swimming pool surrounded by healthy young adults, any one of whom is capable of rescuing me without so much as straining a single fiber of Schwarzeneggeroid muscle, I have more than once felt the dread certainty of imminent drowning; it has been exploded into my brain by the simple realization that I have wandered a few inches beyond my depth.

With an American colleague and a half dozen faculty members of the Hunan Medical University near the south-central Chinese city of Changsha, I was leaving the site of an elaborate banquet (during which my entire alcoholic intake had consisted of one bottle of Tsingtao beer consumed during the early portion of a two-hour meal), chatting and strolling along a curving walkway that stretched a short distance through what appeared to be a shallow reflecting pool. I was fully dressed, and carried a partially filled carry-on bag slung over one shoulder. Having been at the guesthouse two years earlier, I was not unfamiliar with the terrain, but I seem not to have taken into account the narrowness of the winding pavement or the virtual absence of outside lighting on that starless night. As I turned partly around in midstride to address a remark to one of my hosts walking behind me, I suddenly found myself with nothing under my right foot. In an instant, I was immersed well over my head in the impenetrably black water, and still sinking. Simultaneously with the flash of realization that I was fully vertical and continuing to go ever deeper, I felt a stunned surprise and a mild but very distant sense of ironic amusement, as though I were involved in some ill-advised and silly stunt that hadn't worked out quite as I had planned. At the same time, I was annoyed with myself for what I immediately recognized—even down there and seemingly headed through a narrow waterway leading through the earth's crust directly back to New Haven—as a bit of clumsiness that might interfere with the successful completion of my mission in Hunan. Most remarkably, there was no sensation of fear and certainly no thought that I might be drowning.

Although I was not aware of it, I must have finally hit bottom and instinctively kicked off from it like an experienced swimmer, because I soon found myself rising straight up, until my head broke the surface. Taking hold of the outstretched hands of my shouting, frightened companions, I clambered out of the pool, using as foot-holds the irregular projecting rocks that formed its sides. The bag was still on my shoulder; all I had lost were a pair of eyeglasses and some of that necessary element of dignity the Chinese call *mianzi*, or "face." For a few moments, I stood there on the walk-way, feeling stupid, embarrassed, and suddenly very chilly.

My deep dip could not have lasted more than a few seconds, and the summoning of endorphins is only another presumption without possibility of proof. But I relate this episode as a personal testimony to a sudden unanticipated circumstance that should have provoked chaotic loss of control, and yet resulted only in a de-tached imprint of calm and quite reasonable observations about the fix I had (literally) fallen into. The element of emotional shock seems to have triggered a stress response that deprived me of the awareness of danger, thereby preventing the panic-stricken disin-tegration that might otherwise have taken place. I was saved, it appears, from the ineffectual flailing of arms and the aspiration of a few quarts of stagnant water, not to mention the virtual cer-tainty of slamming my thrashing head against the jagged rocks that were only inches away.

My brief moments of peril were hardly of the magnitude of sensory assault visited on a Montaigne or a Livingstone, and I am not so insensitive as to compare them with the tragic events that befell little Katie Mason. And yet, except for a vast difference in degree, they all seem to illustrate the same phenomenon—apparent tranquillity instead of terror, and resignation in the place of self-defeating struggle. Many have pondered the reasons these things should be so, and the answers are spread out over a philosophical terrain as wide as the distance between spiritualism and science. Whatever the source, humankind and many animals often seem to be protected at the instant when sudden death approaches—protected not only from the horror of death itself but from certain kinds of counterproductive actions that might ensure it or extend its anguish.

Here I approach hazardous, but unavoidable, territory. The phenomenon called (and frequently capitalized for emphasis) the Near-Death Experience has been much discussed of late. No sensible observer can discount the many tales from the almost-beyond that have been collected by reliable investigators interviewing credible survivors. Those seeking to interpret their findings on a reasonably scientific basis have invoked a variety of possible causes, from the psychiatric to the biochemical. Others seek clarification in religious faith or parapsychology, while still others accept the experiences at face value, believing them to be not only real but, in fact, the first stages of entrance to a blissful afterlife, virtually always in heaven or its equivalent.

The psychologist Kenneth Ring has interviewed 102 survivors of life-threatening injuries or illnesses. Forty-nine of them met his criteria for deep or moderate near-death experience, and fifty-three were found to be what he called "nonexperiencers." The great majority of illness interviewees had suffered a sudden episode, such as a coronary infarction or a hemorrhage. Dr. Ring found certain basic sequential elements among his positive responders: peace and the sense of well-being; body separation; entering the darkness; seeing the light; entering the light. Other, less universal characteristics include a review of one's life, an encounter with a "presence," meeting deceased loved ones, and making the decision to return. Some of Dr. Ring's patients were medically so far gone that they were thought to be clinically dead, but most were not yet at that point, being merely life-endangered.

I have no more basis for interpreting this so-called Lazarus syndrome than do most others who have pondered it, but I would like to be a little more respectful of the observed facts than have some of the more wishful among them, especially those who go so far as to call the object of their lucubrations the *After*-Death Experience. To do that, I find it helpful to look at the phenomenon's possible biological consequences—what might its function be, and how is it beneficial in the preservation of individuals and the species?

I believe that the near-death experience is the result of a few million years of biological evolution, and that it has a life-preserving function for the species. Very likely, it is similar in nature to the process described in the foregoing pages. The fact that it seems in some few instances to occur even when "death" has been prolonged or relatively stress-free does not alter my expectation that it will one day be proven to be driven, if not specifically by endorphins, then by some similar biochemical mechanism. I would not be surprised if some of the other elements that have been thought to be possible causes do prove to play a role, such as the psychological defense mechanism called depersonalization, the hallucinatory effect of terror, seizures originating in the temporal lobes of the brain, and insufficient cerebral oxygenation. In turn, release of biochemical agents may, in fact, be the consequence or the instigator of one or several of such processes. In the few possible cases where the phenomenon takes place during the lingering death of terminal patients, of course, other factors may play a role, such as injected narcotics or toxic materials produced by the illness itself.

Like so many other biochemical explanations of obscure, seemingly mystical phenomena, this one has no argument with the religious among us. I am neither the first to wonder about the mysterious ways in which God is thought to work His inscrutable will nor the source of the rumor that He may use chemicals to do it. As a confirmed skeptic, I am bound by the conviction that we must not only question all things but be willing to believe that all things are possible. But while the true skeptic can exist happily in a permanent state of agnosticism, some of us have a wish to be convinced. Something within my rational soul does rebel at the invoking of parapsychology, but certainly not of God. Nothing would please me more than proof of His existence, and of a blissful afterlife, too. Unfortunately, I see no evidence for it in the near-death experience.

I do not doubt the existence of the near-death phenomenon and of the equanimity that is sometimes felt when mortality suddenly threatens. Nevertheless, I question the frequency of its occurrence in circumstances other than those that are sudden. Certainly, the comfort and peace, and especially the conscious serenity, of final lingering days on earth have been vastly overestimated by many commentators; we are not well served by being lulled into unjustified expectations.

VII

Accidents, Suicide, and Euthanasia

In a frequently quoted 1904 address at Harvard, the Ingersoll Lecture on the Immortality of Man, William Osler spoke of having in his possession what he described as the deathbed records of about five hundred people, "studied particularly with reference to the modes of death and the sensations of the dying." The case histories of only ninety, said Osler, showed evidence of pain or distress. Of the five hundred, "the great majority gave no sign one way or the other; like their birth, their death was a sleep and a forgetting." The dying are, in Osler's description, "wandering, but uncertain, generally unconscious and unconcerned." Lewis Thomas goes even further: "I have seen agony in death only once, in a patient with rabies." At the time of their utterances, both Osler and Thomas were (Thomas still is) among the most highly respected medical savants of their eras.

And yet I am puzzled. I have seen too many people die in suffering, too many families tormented by the deathwatch they must helplessly keep, to think that my own clinical observation is somehow a misapprehension of reality. The last weeks and days of far more of my patients than Osler's one in five have been overfull with a plethora of purgatory, and I have been there to see it. Perhaps Thomas's different vision is the result of his having spent by far the greater part of his career as a researcher in laboratories; perhaps Osler's interpretation of the five hundred records is reflective of his well-known optimism that the world is really a much better place than we take it to be and of his zeal as a universal mentor to transmit that rosy philosophy. Whatever it is that may have motivated these two most humane of medical scholars to their statements, I must, as we say in such awkward instances of seeming to doubt our own household gods, respectfully disagree.

But then again, I may not disagree at all. Or maybe it is Osler and Thomas who disagree with their own idealizations but have just not been willing to say so. It seems quite possible that they have both begged the question, and done it artfully. In describing what they purport to be the absence of agony in dying, they have conveniently ignored the events that immediately precede the final days or hours of which they speak so reassuringly. With deep sedation or the blessed respite of terminal coma that comes to some at the end of a difficult struggle, the actual hour when the heart stops is indeed often tranquil. Many do, in this way, avoid a tormented passage; but many others are in physical and mental distress till nearly the last moment, or even *at* the last moment. There is a nice Victorian reticence in denying the probability of a miserable prelude to mortality, and it is what everyone wants to hear. But if peace and dignity are what we delude ourselves to expect, most of us will die wondering what we, or our doctors, have done wrong.

For Osler himself, the very end did prove to be peaceful. It was bought, nevertheless, at the cost of much suffering, which even his perpetually cheerful nature was not able to overcome. His final illness lasted two bedridden months, beginning with symptoms thought to be caused by a cold, then influenza, then pneumonia. Though he bore bravely the high fevers and agonizing bouts of uncontrolled coughing, it was sometimes difficult to reassure his wife and worried friends that his optimism was not wilting. Late in his sickness, he wrote in a letter to his former secretary, "I have been having a devil of a time—in bed six weeks!—a paroxysmal bronchitis, not in either of your books! practically no physical signs; cough constant, short couples and then bouts, as bad as whooping cough. . . . Then the other night, eleven o'clock acute pleurisy. A stab and then fireworks, pain on coughing and deep breath, but 12 hours later a bout arrived which ripped all pleural attachments to smithereens, and with it the pain. . . . All bronchial therapy is futile—there is nothing my good doctors have not made me try, but the only things of any service whatever in checking the cough have been opiates—a good drink out of the paregoric bottle or a hypodermic of morphine."

By then, even a spirit as bright as Osler's was flagging, as well as losing its ability to convey optimism to those around him. He had undergone two operations under general anesthesia to drain the pus accumulating in his chest, and each had left him only briefly improved. His torment made him long for the death he had described fifteen years earlier, in which he would be "generally unconscious and unconcerned." Toward the end, the courageous Osler admitted both the difficulty of his passage and his longing for the suffering to end: "The confounded thing drags on in an unpleasant way—and in one's 71st year the harbour is not far off."

Two weeks later, Osler was dead, at the age of seventy. He had lived the threescore years and ten promised by the Book of Psalms. His pneumonia had not been the "acute, short, not often painful illness" that he had long ago described, and it had certainly not fulfilled its function as "the friend of the aged," since he almost certainly would have had many healthy years ahead of him had he not been felled by it. And thus his dying betrayed his expectations, as it will for most of us.

By and large, dying is a messy business. Though many people do become "unconscious and unconcerned" by lapsing or being put into a state of coma or semiawareness; though some lucky others are indeed blessed with a remarkably peaceful and even conscious passage at the end of a difficult illness; though many thousands each year quite literally drop dead without more than a moment's discomfort; though victims of sudden trauma and death are sometimes granted the gift of release from terror-filled pain—conceding all of these eventualities—far, far fewer than one in five of those who die each day are the beneficiaries of such easy circumstances. And even for those who do achieve a measure of serenity during separation, the period of days or weeks preceding the decline of full awareness is frequently glutted with mental suffering and physical distress.

Too often, patients and their families cherish expectations that cannot be met, with the result that death is made all the more difficult by frustration and disappointment with the performance of a medical community that may be able to do no better—or, worse yet, does no better because it continues to fight long after defeat has become inevitable. In the anticipation that the great majority of people die peacefully in any event, treatment decisions are sometimes made near the end of life that propel a dying person willy-nilly into a series of worsening miseries from which there is no extrication—surgery of questionable benefit and high complication rate, chemotherapy with severe side effects and uncertain response, and prolonged periods of intensive care beyond the point of futility. Better to know what dying is like, and better to make choices that are most likely to avert the worst of it. What cannot be averted can usually at least be mitigated.

No matter the degree to which a man thinks he has convinced himself that the process of dying is not to be dreaded, he will yet approach his final illness with dread. A realistic sense of what is to be expected serves as a defense against the unrestrained conjurings of warrantless fear and the terror that one is somehow not doing things right. Each disease is a distinctive process—it carries its own particular kind of destructive work within a framework of highly specific patterns. When we are familiar with the patterns of the illness that afflicts us, we disarm our imaginings. Accurate knowledge of how a disease kills serves to free us from unnecessary terrors of what we might be fated to endure when we die. We may thus be better prepared to recognize the stations at which it is appropriate to ask for relief, or perhaps to begin contemplating whether to end the journey altogether.

There is a kind of dying for which very little or no preparation is possible, and perhaps not advisable. Death by violence is by and large the province of the young. Even when forewarned, youth does not heed the counsel that advises an acquaintance with the avenues leading toward the grave. Neither is youth influenced by statistics—trauma, defined as a physical injury or wound, is the leading cause of death for all persons below the age of forty-four in the United States. It kills approximately 150,000 Americans each year, of all ages; an additional 400,000 are permanently disabled. Sixty percent of the mortality occurs within the first twenty-four hours after injury.

Not surprisingly, our nation's leading source of trauma is automotive. Some 35 percent of major injuries are sustained by automobile occupants and another 7 percent by motorcyclists. The vehicular injuries have at least the virtue of being unintentional in the vast majority of cases. Not so with gunshot wounds (which account for 10 percent of all major trauma) and stabbings (which add almost an equal number). Pedestrian accidents make up 7 to 8 percent, and an additional 17 percent result from falls, which so often involve the very old and the very young. The remaining 15 percent of major traumas arise from a variety of sources, including industrial accidents, bicycle crashes, and an assortment of suicide injuries.

On a late summer day in 1899, a sixty-eight-year-old real estate broker, ironically bearing the name Henry Bliss, stepped off a trolley car in New York City and was killed by a passing automobile, thereby acquiring the dubious distinction of becoming our country's first automotive traffic fatality. Since then, almost 3 million people have died of motor vehicle injuries. The most important contributing cause in those deaths (their traveling companion, so to speak) has been alcohol. Alcohol is a factor in approximately 50 percent of motor vehicle deaths in the United States. One-third of those who have died were victims of someone else's drinking.

Having argued that individual death is of necessity an integral component in the pattern of biological continuity, I add here the self-evident wisdom that nature requires no help. Her own cellular manipulations render unnecessary and ultimately counterproductive our killing of vast numbers of each other, and of ourselves. Trauma robs the species of its progeny and violates the orderly cycle of renewal and improvement. The traumatic death of a human being serves no useful purpose. It is as tragic to the species as to the family left behind.

How ironic it is, then, that so little of our society's biomedical effort is focused on the prevention and treatment of injuries. Only recently has violence been recognized as a major public health problem in the United States—that the number of deaths due to firearms in our country is, per capita, seven times the figure for the United Kingdom; that the frequency of suicide, the most grievous face of violence, has doubled among children and adolescents in the past thirty years, an increase due almost completely to firearms. Suicide is now the third-leading cause of death in those young age groups.

There are those who argue persuasively that the figures for suicide are much too low; they do not include that insidious form of gradually self-destructive behavior some call "chronic habitual suicide": drugs, alcohol, unsafe driving, dangerous sexual habits, gang membership, and the other ways youth may defy the norms of society. Chronic habitual suicide limits not only the quantity of life but its quality as well. It deprives the rest of us of the talents, the passion, and therefore the societal contributions that might have been made by the unfulfilled lives we are losing, often long before we have lost them. Such losses are immeasurable, and they slowly eat away at the edges of our civilization's fabric.

The term *trimodal* has been applied to the time sequence of traumatic dying: immediate, early, and late deaths. An "immediate death" takes place within minutes of the injury. It includes more than half of all traumatic fatalities and is always the result of injury to the brain, the spinal cord, the heart, or a major blood vessel. The physiological process is either massive brain damage or exsanguination.

"Early death" takes place within the first few hours. The usual cause is injury to the head, the lungs, or the abdominal organs, with bleeding in those regions. Death may be due to brain injury, blood loss, or interference with breathing. Regardless of interval, in fact, about a third of all trauma deaths are due to brain damage and another third to bleeding. Although "immediate deaths" are beyond medical intervention, the lives of many patients who fall into the "early" category can be saved by prompt treatment. It is here that rapid transportation, well-trained trauma teams, and battle-ready emergency rooms make the critical difference. It has been estimated that 25,000 Americans die each year because such resources are not universally available. An example of the effectiveness of a quick delivery system is to be found in the lessons of this nation's armed conflicts. In each of our last four major wars, an incremental change in medical know-how was accompanied by a decremental change in evacuation time. The result was a pattern of vastly improving mortality statistics from one war to the next.

"Late death" refers to those people who die days or weeks after the injury. Approximately 80 percent of those mortalities are caused by the complications of infection and failure of the lungs, kidneys, and liver. These people survive the initial blood loss or head trauma but often have sustained injuries to other organs, such as a perforated intestine, a ruptured spleen or liver, or perhaps a blunt injury to the lung. Not infrequently, surgery is required to stop bleeding, prevent peritonitis, or repair a damaged organ, perhaps removing it in the process. Many of these people, instead of recovering uneventfully, begin within a few days to develop fever, high white blood cell counts, and a tendency for some of their circulating blood volume to pool in inappropriate parts of the body, such as the blood vessels of the intestine, and thus be lost to the general circulation. All of these developments are characteristic of widespread infection, or sepsis, which becomes increasingly resistant to antibiotic and other drug treatment.

If the origin of the sepsis is an abscess or infected postoperative incision, surgical drainage will usually reverse the damage and allow the patient to recover. In many people, however, no drainable abscess can be found, so the symptoms progress. By the end of the first postinjury week, respiratory failure begins to appear in the form of pulmonary edema and pneumonialike processes, resulting in decreased oxygenation of the blood. The lung is one of the first targets of sepsis, but it is soon followed by the liver and the kidney. The entire evolving syndrome is thought to represent an inflammatory response to the presence in the blood of a variety of microbial and other invaders that generate toxic substances. These invaders may be bacteria, viruses, fungi, or even microscopic bits of dead tissue. The microbes, if they can be identified, are often found to originate in the urinary system, with the respiratory and gastrointestinal tracts following in frequency. In many cases, surgical wounds and skin are the sites of origin. In response to the presence of the circulating toxins, the lung and other organs seem to create and release certain chemical substances that have a deleterious effect on blood vessels, organs, and even cells, including the elements of the blood. The tissue cells become incapable of extracting sufficient oxygen from hemoglobin at about the same time that less hemoglobin is being brought to them by the reduced circulation. These events so much resemble the classical picture of cardiogenic or hypovolemic shock that their total effect is called septic shock. If septic shock does not respond to treatment, the vital organs fail one after the other.

The occurrence of septic shock is not restricted to subjects of trauma. It is seen in a variety of illnesses in which a patient's defense mechanisms have become impaired. Not infrequently, in fact, it is the terminal event in such a spectrum of conditions as diabetes, cancer, pancreatitis, cirrhosis, and extensive burns, overwhelming its victims with a mortality rate in the range of 40 to 60 percent. Septic shock is the leading immediate cause of death in intensive care units in the United States, accounting for 100,000 to 200,000 deaths each year.

Once the lung has lost some of its ability to oxygenate the blood and the circulation is impaired by a generally depressed myocardium and pooling in the vessels of the gut, several organs begin to demonstrate the effects of the decreased nourishment. Cerebral function dwindles. The liver loses part of its ability to make some of the compounds the body needs and destroy those it does not. The liver failure compounds a concomitant depression of the immune system and the lessened production of infection-fighting substances. At the same time, the decreased blood flow to the kidney prevents proper filtering and results in an inadequate urinary output and gradually worsening uremia, which is a backup of poisonous products in the blood.

All of this may be complicated by the destruction of cells lining the stomach and intestine, with resulting ulcerations and bleeding. Shock, kidney failure, and gastrointestinal bleeding are often the final events in people who die from the syndrome of posttraumatic failure of multiple organs. Stated another way, multiple organ failure is the end point of sepsis, and therefore the common end point for many patients whose primary process may be trauma or one of the more "natural" diseases of mankind. All the syndrome's characteristics seem to be caused by the effects of the toxins on various organ systems of the body. The ultimate outcome for any individual patient is related to the number of organs that cannot withstand the assault. If three are involved, the mortality is close to 100 percent.

The playing out of the entire process usually takes two to three weeks, and sometimes longer. One of my patients, whose sepsis was the result of pancreatitis, lingered for months as all of us—surgeon, consulting physicians, anesthesiologists, resident staff, nurses, and technicians—called upon every diagnostic and therapeutic technique available in our university medical center to hold back the oncoming tidal wave of multiple organ failure, all to no avail.

The ordeal of patients who die of septic shock is indescribably difficult to watch. The unfolding of the ultimately lethal events follows a predictable pattern. First, there are the fever, rapid pulse, and respiratory distress, or at least some evidence of inadequate oxygenation found when the blood is analyzed. An endotracheal tube will be placed to aid the compromised respiration, but it soon becomes evident that no substantial benefit results. If the patient is not already sedated, his level of consciousness is beginning to fluctuate on its own. CT scans, ultrasounds, numerous blood analyses, and multiple cultures are done, all in an effort to find some remediable source of infection, often in vain. Consultants in groups converge around the cubicle, tapping and talking, and in general contributing to the increasing air of uncertainty. The patient is shuttled back and forth between the intensive care unit and the X-ray department as one or another imaging technique is called upon to seek out a pocket of pus or a locus of inflammation. Every transfer from bed to gurney and back becomes a logistical exercise in disentanglement of lines and wires. The spirits and plans of family and medical team change with each new set of laboratory reports, but only the good ones are shared with the anxious person in the bed, providing that individual can still fully comprehend their meaning. Antibiotics are started, changed, stopped in the hope of some treatable germ appearing in the bloodstream, and then restarted; in only about 50 percent of victims of multiple organ failure will a study of the blood yield microbes that will grow in a laboratory culture.

Various alterations in the blood elements appear, and the clotting mechanism may be inhibited, even to the point of spontaneous bleeding. The liver failure sometimes produces jaundice just as the kidney is showing its first serious evidence of progressive deterioration. Dialysis may be tried as a delaying action if there is still some hope of turning things around. By now, if not before, the anguished patient, providing he can still organize his thoughts, has begun to wonder whether enough can be done *for* him to justify what is being done *to* him. Although he cannot know it, his doctors are starting to wonder the same thing.

And yet everyone continues on, because the battle is not yet lost. But all this time, something unnoticed has been happening—despite the best of intentions, the staff members have begun to separate themselves from the man whose life they are fighting to save. A process of depersonalization has set in. The patient is every day less a human being and more a complicated challenge in intensive care, testing the genius of some of the most brilliantly aggressive of the hospital's clinical warriors. To most of the nurses and a few of the doctors who knew him before his slide into sepsis, there remains some of the person he was (or may have been), but to the consulting superspecialists who titrate the remaining molecular evidences of his dwindled vitality, he is a case, and a fascinating one at that. Doctors thirty years his junior call him by his first name. Better that, than to be called by the name of a disease or the number of a bed.

If the dying man has some luck left to him, he is by this time no longer aware of the drama in which he is the principal actor. He has gone from obtundation to minimal responsiveness or even coma, sometimes spontaneously as his organs fail and sometimes aided by narcotics and other medications. His family has gone from worry, to despair, and finally to hopelessness.

Not only the family but also the nurses and those doctors who have been with the dying man from the beginning gradually become affected by the heat of the crucible at the center of their losing campaign. They begin to question the very process by which they and the swarming consultants make treatment decisions or choose to pursue, with increasing desperation, yet another unpromising diagnostic clue. They torment themselves with the increasingly unavoidable perception that they are magnifying the suffering of a fellow human being in order to keep alive the slim hope of recovery; the most self-scrutinizing of the physicians confront that part of their motivation which is the excitement of solving the riddle and snatching up a glorious last-minute victory when the game seems all but unwinnable.

Their separation from the patient brings some of the members of the treatment team gradually closer to the family, as though a transfer of empathy takes place over the long weeks of the vigil. Especially near the end, the comfort that can no longer be perceived by the dying is bestowed upon those who have already begun to mourn. Rarely are there last words in intensive care units—whatever consolation is to be found must come from the warm embrace of a nurse, or the solace of a doctor's words.

Finally, even those who have been unable to let go—even they—feel the relief that comes with the end of the long suffering. I have seen veteran nurses weep openly when an ICU patient dies; I have seen middle-aged surgeons turn their faces away so that young colleagues might not notice the tears. More than once, my voice, and my spirit, too, have cracked before I could utter the words that had to be said.

Of course, such scenes are not restricted to ICUs—they occur also in the general wards and in emergency rooms. Premature death by disease or unprovoked violence can be viewed dispassionately by only very few in the legions of those who care for the sick. But when the premature death is the result of self-destruction, it evokes a mood quite different from the aftermath of ordinary dying—that mood is not dispassion. In a book about the ways of death, the very word *suicide* appears as a discomfiting tangent. We seem to separate ourselves from the subject of self-murder in the same way that the suicide feels himself separated from the rest of us when he contemplates the fate he is about to choose. Alienated and alone, he is drawn to the grave because there seems no other place to go. For those left out and left behind, it is impossible to make sense of the thing.

I have seen my own attitude toward self-destruction reflected in the response of my eldest child. My wife and I had driven one hundred miles to the city where she was a college senior, because we both agreed that we should be with her when she heard the shocking news that one of her most admired friends had killed herself. As gently as we could, and at first without any of the few details available to us, we told our daughter what had happened. It was I who spoke, and I said it all in two or three short sentences. When I was finished, she stared at us unbelievingly for a moment as the tears began overflowing onto her suddenly flushed cheeks. And then, in an uncontrolled paroxysm of rage and loss, she burst out, "That stupid kid! How could she do such a thing?" And that was, after all, the point. How could she do it to her friends and to her family and to the rest of those who needed her? How could such a smart kid commit such a dumb act and be lost to us? There is no place for this kind of thing in an ordered world—it should never happen. Why, without asking any of us, would this beloved young woman just go ahead and take herself away?

Such things seem inexplicable to those who have known the suicide. But for the uninvolved medical personnel who first view the corpse, there is another factor to consider, which hinders compassion. Something about acute self-destruction is so puzzling to the vibrant mind of a man or woman whose life is devoted to fighting disease that it tends to diminish or even obliterate empathy. Medical bystanders, whether bewildered and frustrated by such an act, or angered by its futility, seem not to be much grieved at the corpse of a suicide. It has been my experience to see exceptions, but they are few. There may be emotional shock, even pity, but rarely the distress that comes with an unchosen death.

Taking one's own life is almost always the wrong thing to do. There are two circumstances, however, in which that may not be so. Those two are the unendurable infirmities of a crippling old age and the final devastations of terminal disease. The nouns are not important in that last sentence—it is the adjectives that cry out for attention, for they are the very crux of the issue and will tolerate no compromise or "well, almosts": *unendurable, crippling, final*, and *terminal*.

During his long lifetime, the great Roman orator Seneca gave much thought to old age:

> I will not relinquish old age if it leaves my better part intact. But if it begins to shake my mind, if it destroys its faculties one by one, if it leaves me not life but breath, I will depart from the putrid or tottering edifice. I will not escape by death from disease so long as it may be healed, and leaves my mind unimpaired. I will not raise my hand against myself on account of pain, for so to die is to be conquered. But I know that if I must suffer without hope of relief, I will depart, not through fear of the pain itself, but because it prevents all for which I would live.

These words are so eminently sensible that few would disagree that suicide would appear to be among the options that the frail elderly should consider as the days grow more difficult, at least those among them who are not barred from doing so by their personal convictions. Perhaps the philosophy expressed by Seneca explains the fact that elderly white males take their own lives at a rate five times the national average. Is theirs not the "rational suicide" so strongly defended in journals of ethics and the op-ed pages of our daily newspapers?

Hardly so. The flaw in Seneca's proposition is a striking example of the error that permeates virtually every one of the publicized discussions of modern-day attitudes toward suicide—a very large proportion of the elderly men and women who kill themselves do it because they suffer from quite remediable depression. With proper medication and therapy, most of them would be relieved of the cloud of oppressive despair that colors all reason gray, would then realize that the edifice topples not quite so much as thought, and that hope of relief is less hopeless than it seemed. I have more than once seen a suicidal old person emerge from depression, and rediscovered thereby a vibrant friend. When such men or women return to a less despondent vision of reality, their loneliness seems to them less stark and their pain more bearable because life has become interesting again and they realize that there are people who need them.

All of this is not to say that there are no situations in which Seneca's words deserve heeding. But should this be so, the Roman's doctrine would then deserve consultation, counsel, and the leavening influence of a long period of mature thought. A decision to end life must be as defensible to those whose respect we seek as it is to ourselves. Only when that criterion has been satisfied should anyone consider the finality of death.

Against such a standard, the suicide of Percy Bridgman was close to being irreproachable. Bridgman was a Harvard professor whose studies in high-pressure physics won him a Nobel Prize in 1946. At the age of seventy-nine and in the final stages of cancer, he continued to work until he could no longer do so. Living at his summer home in Randolph, New Hampshire, he completed the index to a seven-volume collection of his scientific works, sent it off to the Harvard University Press, and then shot himself on August 20, 1961, leaving a suicide note in which he summed up a controversy that has since embroiled an entire world of medical ethics: "It is not decent for Society to make a man do this to himself. Probably, this is the last day I will be able to do it myself."

When he died, Bridgman seemed absolutely clear in his mind

that he was making the right choice. He worked right up to the final day, tied up loose ends, and carried out his plan. I'm not certain how much consideration he gave to consulting others, but his decision had certainly not been kept a secret from friends and colleagues, because there is ample evidence of his having at least informed some of them in advance. He had become so sick that he felt it doubtful that he would much longer be capable of mustering up the strength to carry out his ironclad resolve.

In his final message, Bridgman deplored the necessity of performing his deed unaided. A colleague reported a conversation in which Bridgman said, "I would like to take advantage of the situation in which I find myself to establish a general principle; namely, that when the ultimate end is as inevitable as it now appears to be, the individual has a right to ask his doctor to end it for him." If a single sentence were needed to epitomize the battle in which we are all now joined, you have just read it.

No contemporary discussion of suicide, at least not one written by a physician, can skirt the issue of the doctor's role in assisting patients toward their mortality. The crucial word in this sentence is *patients*—not just people, but *patients*, specifically the patients of the doctor who contemplates the assisting. The guild of Hippocrates should not develop a new specialty of accoucheurs to the grave so that conscience-stricken oncologists, surgeons, and other physicians may refer to others those who wish to exit the planet. On the other hand, any degree of debate about physicians' participation should be welcomed if it will bring out into the open a muted practice that has existed since Aesculapius was in swaddling clothes.

Suicide, especially this newly debated form, has become fashionable lately. In centuries long past, those who took their own lives were at best considered to have committed a felony against themselves; at worst, their crime was viewed as a mortal sin. Both attitudes are implicit in the words of Immanuel Kant: "Suicide is not abominable because God forbids it; God forbids it because it is abominable."

But things are different today; we have a new wrinkle on suicide, aided and perhaps encouraged by self-styled consultants on the limits of human suffering. We read in our tabloids and glossy magazines that the actions of the deceased are, under certain sanctioned circumstances, celebrated with tributes such as are usually reserved for New Age heroes, which a few of them seem to have become. As for the pop cultural icons, medical and otherwise, who assist them—we are treated to the spectacle of those publicized peddlers of death willingly expounding their philosophies on TV talk shows. They extol their own selflessness even as the judicial system seeks to prosecute them.

In 1988, there appeared in the *Journal of the American Medical Association* an account by a young gynecologist-in-training who, in the wee hours of one night, murdered—*murder* is the only word for it—a cancer-ridden twenty-year-old woman because it pleased him to interpret her plea for relief as a plea for death that only he could grant. His method was to inject a dose of intravenous morphine of at least twice the recommended strength and then to stand by until her breathing "became irregular, then ceased." The fact that the self-appointed deliverer had never seen his victim before did not deter him from not only carrying out but actually publishing the details of his misconceived mission of mercy, saturated with the implicit fulsome certainty of his wisdom. Hippocrates winced, and his living heirs wept in spirit.

Though American doctors quickly reached a condemning consensus about the behavior of the young gynecologist, they responded very differently three years later in a case of quite another sort. Writing in the *New England Journal of Medicine*, an internist from Rochester, New York, described a patient he identified only as Diane, whose suicide he knowingly facilitated by prescribing the barbiturates she requested. Diane, the mother of a college-age son, had been Dr. Timothy Quill's patient for a long time. Three and a half years earlier, he had diagnosed a particularly severe form of leukemia, and her disease had progressed to the point where "bone pain, weakness, fatigue, and fevers began to dominate her life."

Rather than agree to chemotherapy that stood little chance of arresting the lethal assault of her cancer, Diane early in her course had made it clear to Dr. Quill and his several consultants that she feared the debilitation of treatment and the loss of control of her body far more than she feared death. Slowly, patiently, with rare compassion and the help of his colleagues, Quill came to accept Diane's decision and the validity of her grounds for making it.

The process by which he gradually recognized that he should help speed her death is exemplary of the humane bond that can exist and be enhanced between a doctor and a competent terminally ill patient who rationally chooses and with consultation confirms that it is the right way to make her quietus. For those whose worldview allows them this option, Dr. Quill's way of dealing with the thorny issue of assent (since then elaborated in a wise and outspoken book published in 1993) may prove to be a reference point on the compass of medical ethics. Physicians like the young gynecologist, and the inventors of suicide machines, too, have a great deal to learn from the Dianes and the Timothy Quills.

Quill and the gynecologist represent the diametrically opposed approaches which dominate discussions of the physician's role in helping patients to die—they are the ideal and the feared. Debates have raged, and I hope will continue to rage, over the stance that should be taken by the medical community and others, and there are many shades of opinion.

In the Netherlands, euthanasia guidelines have been drawn up by common consensus, allowing competent and fully informed patients to have death administered in carefully regulated circumstances. The usual method is for the physician to induce deep sleep with barbiturates and then to inject a muscle-paralyzing drug to cause cessation of breathing. The Dutch Reformed Church has adopted a policy, described in its publication *Euthanasie en Pastoraat*—"Euthanasia and the Ministry"—that does not obstruct the voluntary ending of life when illness makes it intolerable. Their very choice of words signifies the churchmen's sensitivity to the difference between this and ordinary suicide, or *zelfmoord*, literally "self-murder." A new term has been introduced to refer to death under circumstances of euthanasia: *zelfdoding*, which might best be translated as "self-deathing."

Although the practice remains technically illegal in the Netherlands, it has not been prosecuted so long as the involved physician stays within the guidelines. These include repeated uncoerced requests to end the severe mental and physical suffering that is the result of incurable disease which has no other prospect for relief. It is required that all alternative options have been exhausted or refused. The number of patients undergoing euthanasia is approximately 2,300 per year in a nation of some 14.5 million people, representing about 1 percent of all deaths. Most frequently, the act is carried out in the patient's home. Interestingly, the great majority of requests are refused by doctors, because they do not meet the criteria.

Involvement is the essence of the thing. Family physicians who make house calls are the primary providers of medical care in the Netherlands. When a terminally ill person requests euthanasia or assistance with suicide, it is not a specialist to whom he is likely to go for counsel, or a death expert. The probability is that doctor and patient will have known each other for years, as did Timothy Quill and Diane, and even then consultation and verification by another physician is mandatory. The length and quality of Quill's relationship with Diane must have been major considerations in the decision of a Rochester grand jury in July 1991 not to indict him.

In the United States and democratic countries in general, the importance of airing differing viewpoints rests not in the probability that a stable consensus will ever be reached but in the recognition that it will not. It is by studying the shades of opinion expressed in such discussions that we become aware of considerations in decision-making that may never have weighed in our soul-searching. Unlike the debates, which certainly belong in the public arena, the decisions themselves will always properly be made in the tiny, impenetrable sphere of personal conscience. And that is exactly as it should be.

Into all of this, an organization called the Hemlock Society has intruded itself. These pages are not the forum in which to critique the problematic way in which this well-meaning self-help group of generally intelligent people has publicly validated the suicide decisions of those who may suffer from impaired judgment. Nor is it my intention to ventilate more than just a bit of my disdain for the misguided way in which the Hemlock Society's founder, Derek Humphry, has represented himself in the limelight of the media during promotion of his ill-advised cookbook of death, *Final Exit*. But no one should make a final judgment on *Final Exit* without being aware of a startling statistic: A 1991 survey conducted by the United States government's Centers for Disease Control found that 27 percent of 11,631 high school students had "thought seriously" about killing themselves in the previous year, and that one in twelve had actually attempted it. More than half a million young Americans are known to try suicide each year, plus an undiscoverable other huge group of those whose attempts are never disclosed.

In a June 1992 letter to the *Journal of the American Medical Association*, two psychiatrists at the Yale Child Study Center advised: "With its lurid examples, explicit instructions, and vigorous advocacy for suicide, *Final Exit* may have an especially pernicious effect on adolescents, who, with their high rate of attempted and completed suicide, appear susceptible to imitative influences and cultural factors that glorify or destigmatize suicide."

Depression, the periodic despondency of the chronically ill, and the death fascination of some segments of our society are not strong enough justifications for teaching people how to murder themselves, to help them do it, or to bestow a blessing on it. No one with impaired powers of judgment is in a position to make a critical decision about ending his or her own life—on that point, there is no disagreement, even among the ethicists who argue most persuasively for the concept that has recently come to be known as "rational suicide." In no way, as Dr. Quill has pointed out, does Derek Humphry's death primer "resolve the profound moral, ethical and personal uncertainties it raises about the meaning of euthanasia and assisted suicide." As with all issues that deal with human life, there is no universal answer, but there should be a universal attitude of tolerance and inquiry. It is perhaps too much to ask that there should also be a universal method of decision-making that is more specifically stated than the guidelines already described. Until a better one is available, Dr. Quill's way—of empathy, unhurried discussion, consultation, questioning, and challenged assumptions—will do just fine.

Though Humphry's philosophy can be condemned, his method cannot. The by-now-well-known technique of swallowing a quantity of sleeping pills just before enclosing one's head in a firmly secured airtight plastic bag does work quite as well as Humphry suggests, even if not by exactly the physiological mechanism he describes. Because the bag is so small, the oxygen is used up quickly, well before the rebreathed carbon dioxide has any significant effect. Rapid cerebral failure ensues, but what really causes death is that a low blood-oxygen level slows the heart quickly to a complete standstill and the arrest of circulation. There may be some symptoms of acute heart failure as the rate of ventricular contraction decreases, but it hardly makes a difference, because dying is so efficiently accomplished. Although one would assume there might be terminal convulsions or vomiting inside the bag, this apparently rarely, if ever, occurs. Dr. Wayne Carver, the chief medical examiner of the state of Connecticut, has seen enough of such suicides to assure me that their faces are neither blue nor swollen. They look, in fact, quite ordinary—just dead.

Each year, some thirty thousand Americans commit suicide, and most of them are young adults. This figure refers, of course, only to those whose deaths can with some certainty be attributed to self-destruction. The stigma that still attaches to suicide is sufficient that families, and the subjects themselves, will often disguise the circumstances. Survivors sometimes appeal to a sympathetic physician to write something else on a death certificate. Elderly males, as indicated earlier, kill themselves at the highest rate per thousand, giving in to the stress of physical illness and loneliness, and being particularly prone to depression.

The great majority of suicides still use the old-fashioned methods of firearms, stabbing, hanging, pills, and gas, or a combination of several. Not infrequently, a poorly planned suicide is botched, especially when attempted by an emotionally distraught individual. In desperation, such people sometimes keep trying until they succeed, resulting in a body being discovered that has been lacerated, shot, and finally poisoned or hanged. When Seneca ultimately did take his own life, it was not by choice but on the order of the emperor Nero. Although one might think his many years of contemplation on the subject had made him something of an expert on its accomplishment, that was not the case—he was a renowned statesman, but he did not know much about the human body. In his determination to make an end of things, he plunged a dagger into the arteries of his arm; when the blood did not come fast enough to suit him, he cut the veins of his legs and knees. That not sufficing, poison was swallowed, also in vain. Finally, records Tacitus, "He was carried into a [heated] bath, with the steam of which he was suffocated."

Barbiturates, a more modern agent of suicide, kill in several ways. The coma they induce is so profound that the upper airway may become obstructed because the head droops into a dangerous position, cutting off the intake of air. That, or the aspiration of vomit, then results in asphyxia. Barbiturates in very high dosage also cause a relaxation of the muscle in arterial walls, allowing the vessels to dilate enough so that blood is lost to the circulation by pooling. In such large amounts, the drugs suppress the contractility of the myocardium and can thus cause cardiac arrest.

In addition to barbiturates, there are several other common pharmacologic agents of dispatch: Heroin, like some of the other intravenous narcotics, kills by causing rapid pulmonary edema, although the mechanism that makes it happen is not known; cyanide inhibits one of the biochemical processes by which cells use oxygen; arsenic damages several organs, but its ultimate way of killing is to produce irregularities of cardiac rhythm, sometimes with coma and convulsions.

When a would-be suicide hooks up one end of a hose to an automobile's exhaust pipe and inhales at the other, he is taking advantage of the affinity that hemoglobin has for carbon monoxide, which it prefers by a factor of 200 to 300 over its life-giving competitor, oxygen. The patient dies because his brain and heart are deprived of an adequate oxygen supply. The color imparted to the blood by the carboxyhemoglobin makes it significantly brighter and paradoxically even more vibrant than its normal state, with the result that the skin and mucous membranes of a person who dies by carbon monoxide have a remarkable cherry-red tinge. The absence of the typical bluish discoloration of asphyxia may deceive those who discover what appears to be a pink-cheeked body in the bloom of health, but dead nevertheless.

Hanging accomplishes much the same thing, but by a mechanism significantly less gentle. The weight of the victim's body provides enough force to tighten the noose and bring about mechanical obstruction of the upper airway. The obstruction is sometimes caused by compression or fracture of the windpipe, but it may also be the result of upward displacement of the base of the tongue, which blocks off the ingress of air. Because the constricting noose cuts off drainage through the jugular and other veins, deoxygenated blood is dammed back up into the tissues of the face and head. The discovery of a grotesquely hanging corpse whose swollen, sometimes bitten tongue protrudes from a bloated blue-gray face with hideously bulging eyes is a nightmarish sight upon which only the most hardened can gaze without revulsion.

In a legal, or judicial, hanging, the executioner attempts to avoid asphyxia, but sometimes he fails. When the knot of the noose is properly positioned just beneath the angle of the condemned man's jaw, the sudden drop of five to seven feet should fracture and dislocate the spinal column at the base of the skull. The spinal cord is thereby torn in two, causing immediate shock and paralysis of respiration. Death, if not instantaneous, is very quick, although the heart may continue to beat for a few minutes.

The sequence of events in suffocation by suicidal hanging is similar to that in all cases of mechanical asphyxia, intentional or not, such as smothering or choking. Nonsuicidal choking is exemplified by the well-known "café coronary," in which a bulky chunk of food suddenly obstructs the windpipe of a diner, often drunk. Made panicky by his inability to take in a breath, the agitated, hypercarbic victim, in a futile attempt to help himself, grabs at his throat and chest as though he is having a heart attack (hence the name café coronary). He will rush toward the bathroom, hoping to vomit up the suffocating plug in his windpipe, because even in his dying moments he remains too embarrassed to do it in front of his gaping fellow diners, who may be sitting there aghast and unable to act. If he is at home and alone, he will probably die, but the Heimlich maneuver may save him if he is in a public place and a bystander can manage it.

If the food plug is not forced out, the process of suffocation continues unchecked. The pulse quickens, the blood pressure rises, and the level of carbon dioxide in the blood increases rapidly to a state called hypercarbia. Hypercarbia produces extreme anxiety, and the decreased oxygen makes the frightened victim appear blue, or cyanotic. He makes increasingly strenuous attempts to pull air past the obstruction, which only serve to wedge the plug even more firmly in place. Just as in a hanging, unconsciousness supervenes, and sometimes convulsions triggered by the unoxygenated and hypercarbic brain. In a short time, the efforts to breathe become weaker and more shallow. The heartbeat becomes irregular, and finally stops.

Drowning is, in essence, a form of asphyxia in which the mouth and nostrils are occluded by water. If the drowning is suicidal, the victim will not resist the inhalation of water, but if accidental, as is usually the case, he will fight it by holding his breath until becoming too exhausted and hypercarbic to continue. At this point, the air passages all the way down into the lung become obstructed by water. If the struggle takes place while the drowning person is thrashing about near the surface, enough air may be sucked in to create a barrier of foam. The foam and water in the airway can set off the vomiting reflex, which adds to the problem by forcing the acid stomach contents up to the mouth, from where they can be aspirated into the windpipe.

If the drowning takes place in fresh water, the water is absorbed into the circulation through the lungs, diluting the blood and upsetting its delicate equilibrium of chemical and physical elements. Red blood cells are destroyed by the imbalance, resulting in the release of large amounts of potassium into the circulation, an element that functions as a cardiac poison by inducing the heart to fibrillate. Should the drowning occur in seawater, the process is virtually the reverse. Water leaves the circulation and enters the alveoli of the lung—the picture produced is that of pulmonary edema. Pulmonary edema may also occur during drowning in a swimming pool, because chlorine acts as a chemical irritant on the lung tissue.

In the struggles of a drowning victim, the aspiration of water is at first delayed and then abetted by one of the body's inherent survival mechanisms. When the first bit of water enters the airway, the larynx reflexly goes into spasm and closes off in an effort to prevent further intake. But within two or three minutes, the decreasing blood oxygen relaxes the spasm and water rushes in. It is this so-called terminal gasp phase that allows the aspiration of so much water that its absorption in a freshwater drowning may account for as much as 50 percent of the blood volume.

A lifeless human body is heavier than water, and the head is its densest part. Accordingly, the corpse of a drowning victim will always sink headfirst to the bottom and remain floating in that position until putrefaction produces enough gas in the tissues to create a buoyancy that makes it rise to the surface. This process takes anywhere from a few days to a few weeks, depending on the temperature and condition of the water. When the body returns, it is difficult for its appalled discoverer to believe that this rotted thing once contained a human spirit and shared nature's life-giving air with the rest of healthy humanity.

Drowning kills almost five thousand people in the United States each year, and alcohol is involved 40 percent of the time. Except in cases of suicide or murder, it almost always occurs under conditions of suddenness, and usually without warning. Nevertheless, the great majority of drowning victims do at least have some sense of its possibility, since it ordinarily takes place when they are in proximity to deep water. The approximately one thousand Americans who yearly suffer lethal electrocution, however, almost never suspect they are about to die, even when they are working around high-tension equipment. By far the most common cause of death following electric shock is ventricular fibrillation caused by the passage of current through the heart. Fibrillation or arrest may also be caused by high-voltage electricity reaching the cardiac center of the brain. If the brain's breathing center is injured, respiratory cessation is the cause of death. Although most lethal electrocutions occur among men who work around high-voltage cables, electrical accidents in the home kill many children and adults each year.

In these various ways, the victims of homicide, suicide, and accidents are deprived of the oxygen supply that maintains existence. This recital of cause and physiological effect hardly exhausts the roll call of soldiers in the squadrons of violent death. Nor does a brief discussion of terminal equanimity, near-death experiences, or assisted suicide more than begin to address the many new issues that have lately been added to the already-lengthy catalog of concern that merits the attention—more than the attention, the scrutiny—not only of philosophers and scientists but of all of us. In matters touching on death, the clinical and the moral are never so far apart that we can look at one without seeing the other.

VIII

A Story of AIDS

"CALL ME ISHMAEL." She smiled at the recollection of that irony, and looked beyond me with wistful eyes into the room where the father of a young family lay dying.

"It was only four months ago, but it's a lifetime, really. I walked into the clinic that day, and there he was, sitting in a cubicle waiting for the great miracle-doctor who was coming to help him. The doctor was me. 'Good morning, Mr. Garcia,' I said, just as bright and breezy as a new interne is supposed to be. And he jumped up, this little Hispanic guy with a great big smile on his face. 'Call me Ishmael' was what he said—imagine it! I guess he never read the book. Melville's Ishmael survived, and mine never had a chance. He'll be dead in a few days, but I'll remember him the rest of my life." She paused; I could tell that the next words were caught on some jagged thing in her throat, because they sounded lacerated when she was finally able to force them out. "He was my first patient with this fucking goddamn disease!"

One crisis after another had taken place since that summertime afternoon when Ishmael* Garcia leaped up from the chair and stuck out his open palm to shake hands with Dr. Mary Defoe, and both of them had vastly changed from what they had been. Though she had seen plenty of AIDS patients while in medical school, Mary never quite realized the full magnitude of individual

*His name was officially Ismail, which is the Spanish form. But since he and others used the English form Ishmael more or less interchangeably with the Spanish, I will also refer to him as Ishmael.

catastrophe until she actually took on the intimidating responsibilities of a newly graduated doctor.

From the sunny July afternoon when he first presented himself to the AIDS clinic until the chilly gray November morning when she was destined to pronounce him dead, Mary Defoe and Ishmael Garcia would be doctor and patient. Whether hospitalized or being followed in the outpatient clinic, he thought of her as his personal physician. From time to time, other internes assumed his care for brief periods when Mary rotated to a different service, but they always found each other again and resumed their journey toward the grim conclusion they both knew lay ahead.

Early in training, most doctors develop relationships with patients that become models on which they will base their responses to sickness and death for the rest of their careers. For Mary Defoe, Ishmael Garcia will surely represent a reawakening of an old image long lost to modern generations of healers—impotence in the face of a plague of death upon the young.

In the calculus of death, no one before 1981 could have factored in HIV, the human immunodeficiency virus. The first hints of its gathering fury struck just at the instant when biomedical science was beginning to offer cautious congratulations to itself on having achieved a state of advancement where the final conquest of infectious disease seemed at last within sight. AIDS not only confounded the microbe hunters; it shook the confidence held by all of us that technology and science can keep humanity safe from the whims of nature. In a very few explosive years, virtually every young doctor in training was treating his or her share of those dying who were meant to live.

Dr. Defoe and I stepped into Ishmael's room—noiselessly, though he was far beyond hearing any sound we might have made. It was more out of respect than necessity that we were so quiet. When a man is dying, the walls of his room enclose a chapel, and it is right to enter it in hushed reverence.

How different this scene from the frenzied drama so often played out during a patient's last moments, as desperate attempts are made to revive him to yet another few weeks or months of waiting for death—and sometimes only hours or days. After the incalculable miseries of Ishmael Garcia's descent into the valley of fever and incoherence, this oblivion was earned; it was fitting that the end, at least, should be undisturbed.

The room's overhead illumination had been turned off and the blinds were closed against the glare of midday autumn sunshine, bathing the entire space in a uniformity of subdued daylight. The unconscious man in the bed had a high fever—the yellowish skin of his forehead glistened against the stark whiteness of his freshly changed pillowcase. Ravaged as he was by the wasting effects of his disease, it could be seen that he had once been very handsome.

I had read Ishmael's chart, and I knew that with his very last breath, the tranquillity would be shattered by a full-scale attempt at resuscitation. In a moment of terror months earlier, he had begged his wife to see to it that the doctors did everything possible to preserve his life—that she not allow them to give up. And now, Carmen could not make herself believe what the AIDS team was telling her: that the possible had become impossible. She clung to that part of her pledge which would destroy the easy exit of an essence in which she devoutly believed—the immortal soul of her husband.

Though Ishmael had been separated from his wife for three years prior to his illness, she was nevertheless his legal next-of-kin, and she spoke for his family. In reality, she spoke only for herself, because Carmen and her husband had together made the unyielding decision to keep the diagnosis to themselves. Neither Ishmael's parents nor his two sisters knew the name of his disease. If they did, they never spoke of it.

When she realized just how sick Ishmael was, Carmen had let him return home. Somehow, she found the strength to put aside his years of unfaithfulness and drug dependence, and even the near-poverty into which his irresponsibility had thrown her and their three daughters. He came back so that she could be his nurse and the only one of his family or friends to share his knowledge of the ultimate end. In spite of everything, he had been a good father, she said, and she owed him this much. For the sake of their three girls and for the recollections of a life that once had been, she permitted her dying husband to return.

In refusing to let him die when his time came, Carmen insisted that she was doing one last kindness for Ishmael—it was, after all, what she believed she had promised him. She refused to discuss with the doctors why she would not listen to their reasoned arguments, and none of them had the heart to press her. They supposed, they told me, that somewhere in the depths of her awareness, Ishmael's obvious devotion to their girls made Carmen feel some unjustified element of guilt about her rejection of her prodigal husband and her obdurate refusal to respond to his sputtering intervals of good behavior and promises of reform. The staff had gone so far as to seek a consultation with the chairman of our hospital's Bioethics Committee, but when they told him that a successful resuscitation might be possible, he would not overrule the dictates of Carmen's heart. In circumstances like this, who knows where wisdom lies?

Ishmael was never alone in that room. The three girls were always with him, a constant presence watching over their adored father through the plastic facing of a three-by-two-foot blown-up photograph standing on the wide windowsill. There they were, three beautiful curly-haired kids in party dresses, smiling out at the world and their father on a day much happier than this. I gestured toward the picture, wordlessly asking Mary a question.

"Yes," she replied, "the two older ones come here almost every day, but Carmen doesn't bring the littlest one. The six-year-old just plays by herself around the foot of the bed—she doesn't really understand. The ten-year-old cries; she stands by her father's bedside every minute she's here, wiping his face and stroking it, and she can't stop crying. I try not to come into the room when they're here—it's more than I can bear."

A Spanish Bible lay at the base of the children's photograph. It was open to chapters 27 through 31 in the Book of Psalms, and several of the verses were marked in various colors of a Hi-Liter. I wrote down the verse numbers on an index card and looked them up when I got home:

> 27:9 Hide not Thy face from me, put not Thy servant away in anger. Thou art my help; reject me not, neither forsake me, O God of my salvation.
>
> 27:10 For though my father and my mother forsake me, the Lord will take care of me.

28:6 Blessed be the Lord, because He has heard the voice of my supplication.

It struck me that Ishmael is Hebrew for "God has heard." The name derives from the words spoken by the Lord when He found Sarah's maidservant Hagar in the wilderness after she fled from the wrath of her mistress: "Behold, thou art with child, and shalt bear a son, and shalt call his name Ishmael, because the Lord hath heard thy affliction." God had found the mother and child by a well, to which He then gave a name bespeaking his recognition of their plight, *Be'er-la-hai-roi*, "The well where the One Who lives has seen."

When the biblical Ishmael was fourteen, God again heard and saw, and this time it was the voice of the lad himself to which He responded, saving him from imminent death in the wilderness and promising to make him a great nation.

To the Ishmael lying in that bed, God seemed not to listen. Neither did He listen, nor did He seem to see. Certainly He did not act, in spite of the torment He observed. In this, Ishmael Garcia was like Job, in the face of whose suffering God was not only at first inactive but silent, too, as though having chosen to be without sight or hearing. If God heard Garcia's entreaties or saw his anguish, He did not change His mind. He never does, in this fucking goddamn disease.

I prefer to believe that God has nothing to do with it. We are witnessing in our time one of those cataclysms of nature that have no meaning, no precedent, and, in spite of many claims to the contrary, no useful metaphor. Many churchmen, too, agree that God plays no role in such things. In their *Euthanasie en Pastoraat* quoted in the previous chapter, the bishops of the Dutch Reformed Church have not hesitated to deal quite specifically with the age-old question of divine involvement in unexplained human suffering: "The natural order of things is not necessarily to be equated with the will of God." Their position is shared by a vast number of Christian and Jewish clergy of various denominations; any less forbearing stance is callous and a further indecency heaped upon people already too sorely tried. Although there is a great deal to be learned from the plague of AIDS, the lessons it teaches lie in the realms of science and society, and certainly not within the purview of religious elucidation. We are dealing not with a punishment but with a crime—one of those random crimes that nature now and then perpetrates on its own creatures. And nature, as Anatole France reminds us, is indifferent; it makes no distinction between good and evil.

There is a good deal more to AIDS than its bare clinical facts disclose. Although such a statement may be made about any disease, how much more so may it be said of this specific plague. But no matter the cultural and societal implications of AIDS, certain of its clinical and scientific manifestations must be understood before the full tragedy unfolds of how it kills its victims. The case of Ishmael Garcia is archetypical.

In February 1990, Garcia had his first positive blood test for HIV. The test was done as part of the evaluation of a nonhealing open sore on his left forearm, which brought him to the medical clinic of the Yale–New Haven Hospital. The infection was almost certainly caused by his intravenous drug habit. Because he felt quite well otherwise, especially when the sore cleared rapidly with a short outpatient course of antibiotics, he never kept any follow-up appointment beyond the one at which he was told his diagnosis. In January 1991, he developed a dry cough that gradually worsened over a period of several weeks. As the cough progressed, a feeling of tightness appeared in Ishmael's chest, aggravated by coughing or a deep inspiration. After more than a month during which things gradually worsened, he began to be frightened by the appearance of two new symptoms: a fever, and shortness of breath brought on by even minor activity. When his breathing difficulty reached the point where it increased with no greater movement than walking around his small furnished room in New Haven's barrio, he knew the time had come to go to the hospital.

In the emergency room, a chest X ray demonstrated that Ishmael's lung fields were diffusely infiltrated with a thin whitish haze, representing the large areas in which infection of some sort was preventing proper aeration. Analysis of arterial blood revealed an abnormally low level of oxygen, reflecting inefficient uptake by the infected lung tissue. When the admitting resident looked into his feverish patient's mouth, he saw the familiar clue exhibited by virtually every new case of AIDS—Ishmael's tongue was coated with the telltale milky white fungus of thrush.

The chest findings were consistent with the form of pneumonia most common in AIDS, caused by a parasite called *Pneumocystis carinii*. Ishmael was admitted to the hospital and the doctors passed a snakelike viewing device called a bronchoscope down into the depths of his windpipe, from where a small sample was taken for culture and the microscopic study that revealed the tightly packed globular structures of *Pneumocystis*. He was given antifungal medication for the thrush and started on a course of a highly specific antibiotic for the pneumonia (called pentamidine), and he gradually recovered. During the hospitalization, Ishmael was found to be anemic and to have a low white blood cell count. Although he insisted that he had been eating well, he was sufficiently malnourished that the protein level in his blood was decreased. On being weighed, he was surprised to find that he had lost 4 of his accustomed 140 pounds. The worst news he was given, however, was something he was not yet able to understand: The marker cell of HIV infection, the T4, or CD4, lymphocyte was found to be at a level of 120 per cubic millimeter of blood, very far below normal.

It is not known whether Ishmael complied with discharge instructions to take his prescribed medications, whose function was to prevent further episodes of the lung infection he had by then learned to call PCP, *Pneumocystis carinii* pneumonia. Most likely he did not, because he returned eleven months later, in January 1992, with similar but even worse symptoms. This time, he complained in addition of severe headache and nausea and seemed somewhat confused. An evaluation of his spinal fluid demonstrated the presence of meningitis caused by a yeastlike organism called *Cryptococcus neoformans*. He was also found to have a bacterial infection of the right ear, but he was too mentally befuddled to be aware of it. His CD4 count was down to 50—HIV's destruction of the immune system was progressing rapidly. Although Ishmael almost succumbed to the combination of three separate forms of infection, skillful management by the Yale–New Haven AIDS team got him through. After three weeks in the hospital, he was able to go back to Carmen and the girls, having amassed a bill of some twelve thousand dollars. Because he had long since lost his health insurance after being fired from his factory job as the result of the drug habit, the costs were assumed by the state of Connecticut.

In early July of 1992, Ishmael, who by that time was scrupulously keeping his appointments at the clinic, developed a large painful abscess in his left armpit, requiring surgical drainage. It was on this visit that he first met Mary Defoe. Over the next few weeks, she supervised his outpatient treatment for sinusitis and another ear infection, while the abscess healed as well.

As Ishmael was recovering from his bacterial illnesses, he once again noted that he was often light-headed and dizzy and sometimes had trouble maintaining his balance. Shortly after the onset of these disturbing symptoms, his memory began increasingly to fail him. Carmen became aware that he did not always comprehend even simple sentences. The symptoms progressed over the next month, to the point where he was confused and lethargic most of the time. In spite of Carmen's gratitude to the physicians, she gave in to his entreaties not to be taken to the emergency room. Both she and Ishmael were afraid of what another hospitalization might mean. He was now losing weight more rapidly, and they knew that, once admitted, he might never come home.

Carmen finally called an ambulance when she awoke one morning and found her husband too far gone to resist. By then, Ishmael was almost in coma, his left arm was twitching uncontrollably, and he barely responded to commands shouted into his ear. From time to time, his entire left side would go into a brief convulsion. A CT scan showed findings most consistent with the type of brain infection caused by a protozoan called *Toxoplasma gondii*, although the blood tests did not confirm that diagnosis. The pictures were striking and consisted of multiple small masses on both sides of the brain. Similar lesions are often found in AIDS patients who have a malignancy called lymphoma, but Ishmael's looked more like those of toxoplasmosis.

At this point, the medical staff decided that even though the diagnosis was not firmly established, it would be safest to begin treatment for *Toxoplasma*, in view of its greater frequency than lymphoma in AIDS patients. When only slight improvement could be discerned after two weeks of drug therapy, Ishmael was taken to the operating room, where the neurosurgeons drilled a small hole in his skull and took a tiny bit of brain for biopsy. Microscopic study of the tissue did not identify the protozoa in the brain, but it did show changes the pathologist believed to be caused by healing of the *Toxoplasma*-induced disease. This encouraged the AIDS team to continue their treatment even in the face of residual uncertainty about the diagnosis. Within a week, however, it became clear that Ishmael's condition was worsening. Because no definite *Toxoplasma* had been identified, those members of the team who had argued against that diagnosis now recommended radiation therapy for a presumed lymphoma of the brain. Before the advent of HIV, brain lymphoma was exceedingly rare, but it is now seen with frequency in AIDS patients.

At first, Ishmael responded to the X-ray treatments by a partial awakening from what had become a profound coma. He even reached the point where he was able to swallow small amounts of custard and pureed foods spooned into his mouth by a nurse or Carmen. But the gains were short-lived. The coma returned, his low-grade fever rose to levels of 102°–103° each day, and he developed a bacterial pneumonia in addition to some other generalized infection whose nature was obscure and, in any case, resistant to treatment. This is how things stood on that November noon when Mary Defoe and I stood at Ishmael's bedside.

Although he was deeply unconscious, Ishmael's face was troubled. Perhaps there was some flickering comprehension of his struggles to move air in and out of those infected lungs, or of the decreasing amounts of oxygen being carried to his gasping tissues. He had become septic, and the entire mechanism of his life was failing. Or maybe his troubled expression had nothing to do with the physical distress of breathless tissues. Possibly, something within him was trying to communicate that he was too exhausted to continue—that he was trying to die but couldn't. And yet, is it really possible that he longed for death? Would not any struggle be worth the anguish, for the chance to see his girls one more time? No one knows why the faces of dying men and women look the way they do—an appearance of discomfort may be as meaningless as an appearance of serenity.

Ishmael's travail ended the next morning. Carmen, feeling the closeness of death, had taken the day off from her job in a New Haven cardboard box factory and sat at his bedside as the seconds grew slowly longer between his respirations, until they stopped altogether. Without being approached again, she had told Mary the night before that there was to be no resuscitation—she could see that her promise to her husband had been kept; everything possible had been done. When Ishmael stopped breathing, she simply stepped outside to inform the nurse who had been sitting with her most of the morning. And then, Carmen did something she had refused over and over again while Ishmael was still alive—she asked to be tested for HIV.

In my section of the United States, the Northeast, AIDS has become the leading killer of men between twenty-five and forty-four—this in a region where deaths among this age group from street violence, drug addiction, and gang warfare are as familiar a part of the urban environment as the poverty and hopelessness that breed them. How can one begin to make sense out of this affliction? No wisdom has yet been discovered, no lesson revealed. AIDS as metaphor, AIDS as allegory, AIDS as symbolism, AIDS as jeremiad, AIDS as a test of mankind's humanity, AIDS as an epitome of universal suffering—it is these kinds of lucubrations that consume the intellectual energies of moralists and littérateurs nowadays, as though something good must at any cost be salvaged from this foul scourge. But even history fails us; analogies with past plagues are always found wanting.

There has never been a disease as devastating as AIDS. My basis for making that statement is less the explosive nature of its appearance and global spread than the appalling pathophysiology of the pestilence. Medical science has never before confronted a microbe that destroys the very cells of the immune system whose job it is to coordinate the body's resistance to it; immunity against a swarming score of secondary invaders is defeated before it has had a chance to mount a defense.

Even the inception of AIDS seems to have been unique. There is now sufficient epidemiological evidence to speculate about the possible origins of the outbreak, and the pathways by which it has achieved its present oppressive hold. The virus is thought by some researchers to have been endemic in a different form among certain Central African primates in which it was not a pathogen and therefore caused no disease. Possibly, the blood of an infected animal may have come into contact with a skin or membrane wound of one or more inhabitants of a local village, who then gradually spread it to others in their immediate surroundings. Basing their work on mathematical models, the proponents of this theory estimate that the first primate-to-human transmission may have taken place as long as a hundred years ago. Because of the sparsity of interactions among communities, the disease spread slowly from its hypothetical village of origin. When cultural patterns began to change after the middle of the twentieth century and people traveled more and more from place to place and became more urbanized, the spread of infection rapidly accelerated. Once a large pool of infected people had come into existence, patterns of international travel carried the virus all over the world. AIDS is a jet-propelled pestilence.

Long before it made its presence manifest by the occurrence of so much as a single identifiable case of AIDS, the virus was being spread among thousands of unsuspecting people. The very first inkling of the new disease came in the form of two brief articles in the June and July 1981 issues of the *Morbidity and Mortality Weekly Report* issued by the Centers for Disease Control (CDC). The articles described the occurrence of two previously extremely rare diseases in a total of forty-one young homosexual men in New York City and California. One of the diseases was PCP, and the other was Kaposi's sarcoma (KS). *Pneumocystis carinii* is not known to cause sickness in people whose immune system is intact. Virtually every one of the cases of PCP reported before this time had occurred in patients with immunity suppressed for the purposes of organ transplantation, or by chemotherapy or starvation, although there were also a few instances on record of congenital immune deficiency. The KS seen in these gay men was of a variety much more aggressive than had heretofore ever been encountered. Of the forty-one patients, those few whose blood was evaluated for T lymphocytes—one of the mainstays of the body's immune system—were found to have conspicuously decreased numbers. Some as-yet-unknown factor had destroyed large numbers of these cells and thereby severely compromised these young men's immunity.

Within a few months, there were several more publications telling of similar cases of what was being given the name gay-related immunodeficiency syndrome, or GRID. At medical meetings, in letters, and over the telephone, infectious disease experts were telling one another about similar patients they were seeing. By December, a deceptively laconic statement in the editorial pages of the *New England Journal of Medicine* had outlined the dimensions of the problem and, in a sensitive and almost prescient way, laid out the framework of the research that needed to be done, as well as the social implications that would have to be addressed:

> This development poses a puzzle that must be solved. Its solution is likely to be interesting and important to many people. Scientists (and the merely curious) will ask, Why this group? What does this tell us about immunity and the genesis of tumors? Students of public health issues will want to put this outbreak into social perspective. Gay associations, which are often active and well informed on pertinent health issues, will want to take measures to educate and protect their members. Humanitarians will simply want to prevent unnecessary death and suffering.

Although the editorialist, Dr. David Durack of Duke University, could not have known it, some 100,000 people worldwide were already infected.

By this time, more than a dozen forms of microbes had been identified from the tissues of diseased young men, and most of them were ones that thrive only in conditions of severely compromised immunity. The part of the immune response affected had been found to be the one dependent on T lymphocytes, and this was supported by the great depletion in numbers of certain of those cells (T4, or CD4, cells) in the blood. Because the depressed immunity provides an opportunity for usually rather benign germs to cause serious trouble, the resultant diseases are called opportunistic infections. When Dr. Durack's editorial appeared, it had already been recognized that "the death rate is fearfully high" and "the only patients . . . who were not homosexual were drug users." The disease was renamed acquired immunodeficiency syndrome, or AIDS.

As noted earlier, the appearance of AIDS, as though from nowhere, was a blow to those members of the public health establishment who had by the mid- to late 1970s convinced themselves that the threat of bacterial and viral disease had become a thing of the past. The present and future challenges to medical science, many were certain, would lie in the conquest of the chronic debilitating conditions such as cancer, heart disease, dementia, stroke, and arthritis. Today, barely a decade and a half later, medicine's purported triumph over infectious disease has become an illusion, while the microbes themselves are winning unforeseen victories. The 1980s brought two new sources of fear—the emergence of drug-resistant strains of bacteria and the advent of AIDS. Both problems will be with us for a long time to come. Dr. Gerald Friedland, the international authority who directs the AIDS unit at Yale, expresses the situation in somber terms that foretell an unending menace: "AIDS is now with us for the duration of human history."

The protests of some AIDS activists notwithstanding, the amount of information that has since then been gathered about the human immunodeficiency virus and the progress made in mounting a defense against its onslaughts are nothing less than an astonishment. *Astonishment*, in fact, is precisely the word used in describing the rapidity of progress by the time of the pandemic's seventh year. In 1988, Lewis Thomas, among whose other outstanding accomplishments has been his role as a pioneer of immunology, wrote this:

> In a long lifetime of looking at biomedical research, I have never seen anything to touch the progress that has already been made in laboratories working on the AIDS virus. Considering that the disease was recognized only seven years ago, and that its agent, HIV, is one of the most complex and baffling organisms on earth, the achievement is an astonishment.

Thomas went on to point out that even at that relatively early time, scientists already knew "more about HIV's structure, molecular composition, behavior and target cells than about those of any other virus in the world."

Not only in the laboratory but in the realm of treatment as well, encouraging signs have appeared that patients are today living longer, their symptom-free periods are expanding, and the level of their comfort is improving. These changes are keeping pace with increased knowledge about routes of worldwide spread, public health measures, and the social and behavioral changes that will be necessary if we are to achieve optimum control over the pandemic.

Much of the progress has been made through the active collaboration of universities, government, and the pharmaceutical industry. Such a troika is a welcome phenomenon in American biomedicine, and its existence owes much to the forceful campaigns conducted by AIDS advocacy groups, at first almost exclusively those within the gay community. Patient pressure groups are a relatively new factor in the equation of biomedical research, but an increasingly powerful one. Due as much to the efforts of the AIDS lobby as to the demands of the doctors, approximately 10 percent of the $9 billion budget of the National Institutes of Health now goes to the study of HIV. The U.S. Food and Drug Administration has been kept under constant fire to relax the strict standards it has painstakingly developed in evaluating experimental drugs. In some ways, this has been to the good; conditional approval has been granted for therapeutic agents that have demonstrated sufficient effectiveness under laboratory conditions. The inherent danger of easing hard-won safeguards, however, must be borne in mind—even in times of plague.

Particularly impressive is the rapid series of early discoveries, beginning almost immediately upon the CDC alert. The fact that several cases of PCP in nonhomosexual IV drug abusers had been reported by the end of 1981 gave rise to the probability that the mode of spread of the new disease was similar to that of hepatitis B, a virus commonly found in that group. It was reasoned that the causative agent being sought must be a virus. This theory was given credence in 1982 by a CDC report that nine of the first group of nineteen patients in the Los Angeles area could be linked through sexual contact with one man, and these nine in turn to forty others who had been diagnosed in ten different cities. The finding established the sexual transmissibility and infectiousness of the disease with a degree of certainty beyond doubt.

By mid-1984, the human immunodeficiency virus had been isolated and demonstrated to be the causative agent of AIDS, and its methods of attacking the immune system were clarified. At the same time, the clinical ravages of the disease process had been characterized and a blood test developed. While this was being accomplished in the laboratory and clinic, studies by public health officials and epidemiologists had elucidated the general form and dimensions of the outbreak.

At first, there was considerable skepticism in the scientific community that any drug would ever be found with the capability of damaging the virus itself. Much of the concern grew out of what was becoming known about the characteristics of the microbe, especially the fact that it survives by integrating itself into the very genetic material (the DNA) of the lymphocytes it attacks. Not only that: HIV was found to have the ability to hide in various cells and tissues where it is not only protected but also difficult to find. Additionally, it fools the body's antibody response by a remarkable bit of trickery: the outer envelope of a virus is made of protein and fatty materials, whereas a bacterium is surrounded primarily by carbohydrate. The body's immune response is kicked off much more readily by protein than by carbohydrate. HIV, however, coats its protein envelope with carbohydrate, becoming, in a sense, a virus in bacterial clothing. This insidious masquerade succeeds in decreasing antibody production. As if all of this was not enough, HIV mutates extensively, allowing it to turn itself into a somewhat different strain of beast should the body's antibody response or a new antiviral drug somehow manage to overcome the obstacles placed before it.

Given all of these challenges, plus the fact that HIV kills off the mainstay of the body's defense by destroying the lymphocytes within which it lives, there was reason for discouragement. Almost in desperation, researchers began to carry out laboratory evaluations of a variety of drugs they thought might conceivably fight the evasive virus. In the face of the reality that HIV's duplicity would prevent the early development of a vaccine to mobilize the body's own immunity, scientists adopted the same approach to fighting AIDS that they had been using to combat bacterial infections: They began to search for pharmaceutical agents that function in the same way as antibiotics, by killing the infectious organism or preventing its reproduction without depending on the immune system as a first line of defense.

Some of the agents tested had been intended for other uses, found to have limited effectiveness, and been put back on the shelf. As more knowledge was gained of the specific characteristics of the virus (especially after HIV became available in a form that could be used in the laboratory, in 1984), it was possible to be more focused in the search for effective compounds. By the late spring of 1985, three hundred drugs had been tested at the National Cancer Institute, and fifteen of them were found to stop the reproduction of HIV in the test tube. The most promising of them was an agent first described as an anticancer drug in 1978, bearing the chemical designation of 3-azido, 3-deoxy-thymidine, or AZT (often called zidovudine). AZT was administered to the first patient on July 3, 1984, and large-scale clinical studies were begun at twelve medical centers in the United States. By September 1986, there was sufficient evidence to show that the drug could decrease the frequency of opportunistic infections and improve the quality of life of AIDS patients, at least until the virus mutates against it. It was the first effective therapy ever to have been found against the particular category of viruses in which HIV belongs, called retroviruses. Although the drug is very expensive and potentially toxic, it soon became the mainstay of treatment directed at HIV. The discovery of AZT's effectiveness encouraged the search for other, similar agents. The first to be identified was dideoxyinosine (ddI, or didanosine), and work has continued.

The development of AZT is only one example of the furious efforts sometimes required to combat HIV at that early time. From the beginning, an amount of information has come forth that is sometimes staggering to nonspecialists.

There are ever-deeper insights into molecular biology, improved methods of surveillance and prevention, constant revision of statistical reporting, increased understanding of the pathology wreaked by opportunistic organisms, and, thankfully, new drugs against those infectious jackals and the viruses they follow after.

It is no easy matter to explain or understand the mechanism by which the many opportunistic invaders lay waste the body of an adult or child with AIDS. Such a baffling array of problems is faced by the HIV-infected and their caregivers that one cannot but contemplate with a sense of awestruck gratitude that so much has already been accomplished. When a doctor of my generation makes hospital rounds with an AIDS team of physicians and nurses, all he can do is figuratively gasp at how much these skilled clinicians know and what a large proportion of it has been learned in so short a time. Every patient on the unit carries a multitude of infections and sometimes one or two cancers; each is receiving four to ten or even more medications, without any certainty of predictable response or toxicity—Ishmael Garcia was on fourteen. Daily, and sometimes more often, new decisions must be made about everyone being treated (my hospital's relatively small AIDS area has forty beds, and they are always full).

As if the immensity of the clinical challenges were not enough, distraught families wait nearby for answers, and for consolation as well; there are reports for the staff to fill out, charts to review, tests to order, students to teach, conferences to attend, and an ever-burgeoning new literature to read and often to contribute to. And always, the most important charge is to care for those dreadfully stricken brothers and sisters of all of us, the sickest of whom are wasted, feverish, swollen, and anemic, their eyes seeking some reassurance and the unspoken promise of relief from their torment, which too often will come only with death. No matter the perseverance and moral strength so many patients muster in the face of lethal certainty, the pitiless process by which they die is dispiriting anew with every reenactment.

IX

The Life of a Virus and the Death of a Man

THE DISCOVERIES THAT were being so rapidly made about the life cycle of the virus provided the background against which to seek points where it might be vulnerable to attack. Viewed simply, a virus is nothing more than a tiny particle of genetic material enclosed within the envelope of protein and fatty materials. Viruses are the smallest known living things, and they carry very little genetic information. Because they can't exist without the help of more complex structures, they must live within cells. Since a virus cannot reproduce (in the case of a virus, scientists prefer the term *replicate*) on its own, like a bacterium, for example, it must get itself inside a cell and take control of the cell's genetic machinery by becoming integrated into it. HIV accomplishes this by a method that is the reverse of the ordinary process by which genetic information is transmitted; for that reason, it is called a retrovirus.

The genetic material of cells is composed of the strands of molecules called deoxyribonucleic acid (DNA); DNA is the repository of genetic information. Under ordinary conditions of reproduction, DNA is copied, or "transcribed," into other molecular strands called ribonucleic acid (RNA), which functions as a template to direct the production of the new cell's proteins. In a retrovirus, however, the genetic material is RNA. The retrovirus also carries an enzyme called reverse transcriptase, which, once the virus enters the host cell, transcribes the RNA into DNA, which is then translated in the usual proper sequence into proteins.

The series of events that takes place when a lymphocyte is infected by HIV is roughly as follows: The virus binds to structures called CD4 receptors on the membrane surrounding the cell; at those points, it sheds its envelope as it is taken into the cell, where its RNA is transcribed to DNA. The DNA then migrates into the lymphocyte's nucleus and inserts itself into the cell's own DNA. For the rest of that lymphocyte's lifetime and the lifetime of its progeny, it remains infected with the virus.

From this point on, every time an infected cell divides, the viral DNA is duplicated along with the cell's own genes and remains as a latent infection. For unknown reasons, at some point the viral DNA dictates the production of new viral RNA and viral proteins; in this way, new viruses are manufactured. They bud off from the lymphocyte's cell membrane, are set free, and then go on to infect more cells. If the process is fast enough, it can kill the lymphocyte that harbors it, which is destroyed as the virus particles burst out. Yet another method of lymphocyte destruction makes use of the fact that certain structures on the surface of the newly budded virus can bind to still-uninfected T cells, with the result that large numbers of the cells fuse together into clumps called syncytia. Because syncytia can no longer function in immunity, the clumping proves to be a very effective way of inactivating many lymphocytes at once.

As noted earlier, the cell attacked by HIV is the T lymphocyte, a white blood cell that has a major role in the body's immune response. Specifically, it is a subset of the T cells called CD4, or T4, lymphocytes (yet another name is the helper T cell) that is victimized. So dominant is the CD4 cell in the overall functioning of the immune system, it has been called its "quarterback."

HIV can thus affect CD4 cells in various ways. It can replicate in them, can lie dormant for long periods of time, and can also kill or inactivate them. It is the enormous depletion of CD4 lymphocytes gradually occurring over time that is the major factor in preventing a patient's immune system from mounting an effective defense against various forms of infections by bacteria, yeasts, fungi, and other microorganisms.

HIV attacks another type of white blood cell as well, called the monocyte, of which as many as 40 percent have the CD4 receptor in their membrane, and can thus take on the virus. Yet one more refuge is the macrophage (literally, the "big eater"), among whose functions is the ingestion and destruction of infectious cellular debris. Unlike the CD4 lymphocyte, neither the macrophage nor the monocyte is destroyed by HIV; they seem to be used as reservoirs and safe houses in which the microbe may lie dormant for long periods of time.

All of the foregoing is but a sketchy outline of the way in which the immune system is gradually laid waste by HIV. Although some have protested the use of military analogies to portray the pathophysiology of disease, AIDS lends itself particularly well to such descriptive comparisons. The process, in fact, is not unlike a gradual buildup of forces, during the later stages of which a prolonged artillery and air bombardment destroys a country's defenses in preparation for a massive land invasion carried out by a large coalition of belligerents, allied together to accomplish total annihilation. The army of microbes that kills the victim of AIDS after HIV has knocked off his CD4 cells includes many different kinds of divisions, and every one of them has its own target and its own lethal mechanism of attack. The most conservative epidemiologists predict that by the year 2000, there will be 20 to 40 million seropositive people on our planet who are under siege or already invaded. Forty to eighty thousand Americans are becoming newly infected each year, and the same number die.

As far as has yet been determined, there are only three ways in which infection can take place: via sexual contact, an exchange of blood (as with a contaminated needle, syringe, or blood products), or transmission from an infected mother to her child in the uterus, at the time of delivery, or even in the breast milk postnatally. HIV has been isolated in the laboratory from blood, semen, vaginal secretions, saliva, breast milk, tears, urine, and spinal fluid, but only blood, semen and breast milk have ever been found to transmit the disease. Since 1985, the banked blood supply has been so carefully screened that the possibility of contracting HIV from transfusion is remote. In the United States and most developed nations, the overwhelming majority of those infected by the sexual route are gay or bisexual men, but in Africa and Haiti, the great predominance is among heterosexuals. Although the number of heterosexually transmitted cases remains low in the West, it is gradually rising, as is the number of infected infants. Approximately one-third of the Americans who become infected each year are intravenous drug abusers, and at least an equal number are gay men. The remainder, most of whom are black and Hispanic women, acquire the disease heterosexually, and their seropositivity explains why two thousand babies are born infected each year.

AIDS is a disease of low contagion. HIV is a very fragile virus—it is not easy to become infected with it. Simple household bleach in a 1:10 dilution kills the virus efficiently, as do alcohol, hydrogen peroxide, and Lysol. Within twenty minutes of being poured on a tabletop and allowed to dry, virus-laden fluid is no longer infective. One need not fear any of the four bugbears (or bug-bearers) so often avoided by the germ-shy: insects, toilet seats, eating utensils, and kissing. Although there are certainly cases thought to have been transmitted by a single sexual encounter, seropositivity usually requires a very high dose of virus or repeated episodes of contact. In the United States, the risk of seropositivity via a chance heterosexual intimacy is real but very small. As reassuring as it may be to contemplate the difficulties that must be overcome by the virus in order to infect us, the sense of security disappears in the face of the grim fact that once infected, we are likely to die. That consideration alone justifies the precautions urged on us by public health authorities.

The virus often shows its hand very soon after entering a new host. Within a month or less, rapid replication causes its concentration in the blood to become extremely high, and it stays that way for about two to four weeks. Although many newly infected people remain without symptoms, others during this period develop low-grade fevers, swollen glands, muscle aches, a rash, and sometimes central nervous system symptoms such as headaches. Because these symptoms are not specific and may also be accompanied by a general sense of fatigue, they are often erroneously attributed to flu or mononucleosis. As this brief syndrome is ending, the first antibodies against HIV begin to appear in the blood; a blood test will detect them, and the patient is henceforth seropositive. Although the short symptomatic period ends, the virus continues to replicate.

Very likely, the brief mononucleosislike syndrome is caused by the first response of the body's immune system to the alarm set off by the massive number of new virus particles that have by then been produced. The body is initially successful, and the number of virus-particles in the blood drops dramatically to low levels. What seems to have happened at this point is a retreat by the remaining microbes into CD4 lymphocytes, lymph nodes, bone marrow, the central nervous system, and spleen, where they lie dormant for years or replicate so slowly that the total low concentration in the blood remains stable. Actually, only 2 to 4 percent of the body's CD4 cells are in the blood. Most likely, those in the lymph nodes, spleen, and marrow are being gradually destroyed during the long dormant period, but the destruction is not reflected in the blood until the end of this time, when the CD4 counts, having remained constant till then, begin to drop dramatically, allowing the multiple secondary infections that characterize AIDS to ensue. At that point, the amount of virus in the blood rises again. The reason for the prolonged period of relative inactivity is unknown, but it does suggest that the body's immune system may be playing some role in mitigating the infection, at least that part of it which is restricted to the blood itself. When the immune system has sufficiently deteriorated, the amount of virus in the lymphocytes and free in the blood increases markedly.

This sequence of events may explain why most HIV-positive people develop swollen lymph glands in the neck and armpits during the early two-to-four-week period of symptoms, which do not recede at the end of it. When that period is over, patients continue to feel well for an average of three to five years or even ten years, by the end of which time examination of the blood will usually reveal that the number of CD4 cells has declined considerably, from a normal count of 800 to 1,200 per cubic millimeter to below 400. This means that 80 to 90 percent of these lymphocytes have been destroyed. On an average of eighteen months later, standard skin tests for allergy begin to show that the immune system is becoming impaired. The CD4 count continues to drop, but patients at this stage of disease still may not have begun to show evidence of clinical illness. In the meantime, the level of virus in the blood is rising and the swollen lymph nodes are slowly being destroyed.

When the CD4 cell count falls below 300, the majority of patients will develop the fungal infections of the tongue or oral cavity called thrush, which presents itself as white patches in those areas. Other infections that may begin to appear when the count is below 200 are herpes around the mouth, anus, and genitals, as well as a severe vaginal infection with the same fungus that caused the thrush. A characteristic finding is a condition called oral hairy leukoplakia (from the Greek *leukos*—meaning "white," and *plakoeis*—meaning "flat") a group of fuzzy-looking vertical white patches standing upright like corrugations along the lateral margins of the tongue. These lesions are due to a virus-induced thickening of the surface layers.

Within a year or two of this time, many patients are beginning to develop opportunistic infections in areas beyond the skin and body openings. By then, the CD4 cell count has usually fallen well below 200 and is dropping rapidly. The entire syndrome of immune deficiency begins to make itself evident as diseases appear that are caused by microbes ordinarily living in perfectly healthy people whose normal physiological defenses prevent trouble. The stage has now been reached at which serious pathology can be caused by any organism that requires intact immunity to combat it. Although people with AIDS are highly susceptible to well-known diseases such as tuberculosis and bacterial pneumonia, they are also set upon by a group of otherwise-unusual sicknesses due to a variety of parasites, fungi, yeasts, viruses, and even bacteria that physicians rarely encountered before the advent of HIV. For some of these organisms, there was no effective treatment until the late 1980s, when the efforts of university laboratories and the pharmaceutical industry were finally rewarded with the development of a group of drugs that have shown varying degrees of clinical success.

Every variety of microbial invader attacking the shattered defenses of the immune-compromised person with AIDS is equipped with its own unique assault weapons and directs its onslaughts against specific objectives. With little remaining CD4 cell resistance to bar their way, the individual divisions and regiments of opportunistic killers devastate the territory that comprises the patient's tissues. Sometimes by exhausting a person's energy and small supply of reserve firepower, and sometimes by knocking out a central structure like the brain, the heart, or the lungs, the swarming bits of infection will have their way. Though the pestilential offensives may be slowed or halted for a while by one or another of the newer pharmaceutical agents, they will always in time resume, if not in one form, then in another. A skirmish may be won here and there, or a battle prevented by timely use of prophylactic drugs, and some months of stability thereby achieved—but the eventual outcome of the struggle is preordained. The determined microbial aggressors will accept nothing less than the unconditional surrender that comes only with the death of their involuntary host.

Although AIDS patients may die of any of a number of pathological processes, a relatively small group of microbes is involved in the vast majority of deaths. Foremost among these is *Pneumocystis carinii*, the first one to be identified at the very outset of the worldwide pestilence. The figures are now declining because of prophylactic medication, but until fairly recently, more than 80 percent of patients had at least one experience of PCP, and many died during an episode, either from the respiratory insufficiency itself or from problems associated with it. Depending on the severity of the onslaught, an individual episode used to kill between 10 and 50 percent of its victims before effective means had been found to combat it. It remains a significant factor in the death of nearly half of all AIDS sufferers, but the percentage continues to decrease.

The symptoms of PCP are essentially those experienced by Ishmael Garcia as his breathing became progressively compromised until he sought treatment. Occasionally, the organism may be found in other parts of the body than the lungs, and in autopsies of patients who die of this infection it is sometimes disseminated throughout virtually every major organ, most particularly the brain, heart, and kidneys.

Like patients with other types of pneumonia, those who die with PCP are asphyxiated by the infected lung's inability to be aerated. As wider areas of tissue become involved, more and more alveoli are destroyed, and a point is reached where arterial oxygen levels cannot be raised in spite of every available means of forcing the gas into the soggy and plugged tissues. The lack of oxygen and the buildup of carbon dioxide damage the brain and finally stop the heart. Sometimes destruction of tissue has been so severe that cavities have been created in the areas of disintegration, very much as in tuberculosis.

The lung is the organ most commonly assaulted by AIDS. Virtually every one of the opportunists, as well as the tumors, looks to the lung as a target. On hospital rounds I have attended, tuberculosis, pus-forming bacteria, the herpeslike cytomegalic virus (CMV), and toxoplasmosis are among the most common problems discussed. Except for the last, they all seek out a home in the respiratory tissue. The incidence of tuberculosis in AIDS patients is some five hundred times what it is in the general population.

Toxoplasmosis is a disease that was at one time so rare that I had difficulty remembering just what it was when I first encountered it in an early AIDS patient. In little more than a decade, it has become a major belligerent in the HIV invasion, and I will never again have to search my memory about its details, so devastating are the things I have seen it do to defenseless people. The organism itself is a protozoan commonly found infecting birds, as well as cats and other small mammals. Most commonly, it is transmitted to humans in inadequately cooked meat or is ingested when food is contaminated with animal feces. *Toxoplasma* lives harmlessly in anywhere from 20 to 70 percent of Americans, its frequency depending on the social and economic group tested. In an immunodeficient patient, however, it manifests itself by fever, pneumonia, enlargement of the liver or spleen, rash, meningitis, encephalitis, and sometimes involvement of the heart or other muscles. Its most common focus of attack in AIDS is the central nervous system, where it can cause fever, headache, neurological deficits, seizures, and mental changes ranging from confusion to deep coma. On CT scans, the infected areas of the brain sometimes so much resemble the lesions of lymphoma that they are differentiated only with difficulty. This was the diagnostic dilemma that caused so much uncertainty in the care of Ishmael Garcia.

It is a rare AIDS patient whose nervous system escapes the pillaging of the disease. Even in the early period of HIV infection, some few people go through a transient period of neurological disabilities, which may sometimes appear even before AIDS itself has supervened; fortunately, this particularly distressing complication is far less common in early than in late stages of HIV disease, when it is more severe and called the AIDS dementia complex. Its eventual effects on cognition, motor function, and behavior can be devastating, but most frequently present initially as simple forgetfulness and loss of concentration. After a while, apathy and withdrawal become common symptoms, while some smaller number of patients complain of headaches or develop seizures. Should these findings not pass off when they occur early in HIV infection, they slowly worsen. In that case or in those far more common patients whose symptoms appear in the AIDS period, intellectual function often declines and difficulty with balance or muscular coordination appears. In the most advanced stages of the complex, patients are severely demented and show little response to their surroundings; they may be paraplegic and suffer tremors or occasional convulsions. These complications exist without any relationship to those processes caused by cerebral toxoplasmosis, lymphoma of the brain, or other opportunistic neurological disabilities such as meningitis caused by the yeastlike fungus cryptococcus. AIDS dementia complex is thought to be due to the virus itself, but its exact cause is unknown, and the cerebral atrophy seen on CT scan and biopsy is unrelated to any other factor. Of the many neurological problems associated with AIDS, this one and toxoplasmosis are the most common. Fortunately, the beneficial effects of AZT have resulted in some decline in its frequency.

Two cousins of the tuberculosis germ share the distinction of being the bacteria most frequently disseminated throughout the body of people with AIDS. *Mycobacterium avium* and *Mycobacterium intracellulare* (MAI), jointly called the *Mycobacterium avium* complex (MAC), are present in about half of AIDS patients when they die, having caused a wide variety of symptoms during life. MAI is now a more frequent cause of death than PCP. Fever, night sweats, weight loss, fatigue, diarrhea, anemia, pain, and jaundice are often attributable to these marauding twins. Although the complex rarely causes death on its own, its wasting effects are major contributors to the general debilitation and malnourishment that further weaken defenses against other invaders.

These are just a few of the manifestations of AIDS. Lengthening the list serves only to name some of the other common problems that beset patients, but it cannot even approach the complete inventory of suffering: the blindness of the retinitis resulting from CMV or *Toxoplasma* infection; the massive diarrhea that can have any of five or six causes, or sometimes none that are identifiable; the meningitis or occasional pneumonia of cryptococcosis; the thrush or swallowing difficulties of candidiasis, and perhaps the slimy wet ooze of its skin lesions; the discomfort of herpes around the anus; the fungal pneumonia or bloodstream seeding of histoplasma; the bacteria typical and atypical; the more than a score of creeping, crawling things with names like *Aspergillus*, *Strongyloides*, *Cryptosporidium*, *Coccidioides*, *Nocardia*—their time has come and they act like looters after a natural disaster, which is exactly what they are. Though they pose no danger to people with normal immunity, every one of them is a bane to those with a depleted store of CD4 lymphocytes.

The heart, kidney, liver, pancreas, and gastrointestinal tract are affected in numerous ways by AIDS, as are the tissues less commonly thought of as specific organs, such as the skin, blood, and even the bones. Rashes, sinusitis, clotting abnormalities, pancreatitis, nausea, vomiting, draining sores and noxious discharges, visual disturbances, pain, gastrointestinal ulceration and bleeding, arthritis, vaginal infections, sore throat, osteomyelitis, infection of heart muscle and valves, kidney and liver abscesses—there are many others. Not enough that this disease depletes and dispirits, many patients feel humiliated by the details of their ordeal.

Kidney and liver function are often affected; there may be conduction or valvular abnormalities of the heart; the digestive tract betrays its owner in any number of ways; the adrenal and pituitary glands sometimes lose their power. When bacterial infection is no longer controllable, the familiar picture of sepsis supervenes. All the while, malnutrition and anemia are further weakening the body's ability to slow down the processes of destruction. The malnutrition is often aggravated by huge protein losses through the damaged kidneys, resulting from a rapidly progressive condition of uncertain cause, called HIV-associated nephropathy (kidney disease). The nephropathy may go on to terminal uremia within three to four months of onset.

Even without direct involvement by infection, the heart in AIDS patients occasionally becomes enlarged and may fail, or it may develop a rhythm irregularity leading to sudden death. The liver, too, is susceptible to attack, not only because of AIDS itself but because so many patients are concomitantly infected with the hepatitis B virus. CMV, MAI, tuberculosis, and several of the fungi have a predilection for the liver. The hapless organ is not only battered by the disease but by attempts to treat it, as drug toxicity affects its performance in many ways. The liver is found to be abnormal in some way or other in 85 percent of autopsied patients.

The entire length of the gastrointestinal tract is a vast twisting tunnel of opportunity for the various predators of AIDS. From the herpes and that wide assortment of ulcerations and infections around and in the mouth to the running sores and problems of continence at the anus, the torment of the final months may be magnified by the involvement of so many structures that it inhibits eating, interferes with digestion, and produces uncontrollable watery diarrhea that not only is a source of constantly recurring distress but also makes it difficult to maintain proper hygiene of the raw areas around the anus and rectum. To imagine extracting a scrap of dignity from this kind of death is beyond the comprehension of most of us. And yet the indignity itself sometimes brings out moments of nobility that overcome for a while the reality of anguish—arising from sources so deep, they can only be marveled at, for they surpass understanding.

An intact immune system is needed not only to resist infection but to inhibit growth of tumors as well. In the absence of an effective defense, certain kinds of malignant processes find a favorable environment in which to make their appearance. HIV has been particularly conducive to one form of cancer previously so rare that I had seen exactly one case, in an elderly Russian immigrant, since my graduation from medical school almost forty years ago. The incidence of this malignancy, Kaposi's sarcoma, has been magnified by a factor of well over a thousand—from 0.2 percent of the general population to more than 20 percent of Americans with AIDS. It is by far the most common tumor seen with this disease, and for as-yet-uncertain reasons, it afflicts a greater percentage of homosexual men (40 to 45 percent) than IV drug abusers (2 to 3 percent) or hemophiliacs (1 percent). These figures reflect only the people in whom the diagnosis is made during life. When autopsies are done, the frequency of KS triples or quadruples, making its presence somewhere in the bodies of gay men even more common.

In 1879, Moritz Kaposi, a professor of dermatology at the University of Vienna Medical School, described an entity he called "multiple pigment sarcoma," consisting of a group of reddish brown or bluish red nodules that originates on the hands and feet and advances along the extremities until reaching the trunk and head. In time, stated his report, the lesions enlarge, ulcerate, and spread to the internal organs. "Fever, bloody diarrhea, haemoptysis [coughing of blood] and marasmus set in at this stage, and are followed by death. At the autopsy, similar nodules are found in large numbers in the lungs, liver, spleen, heart and intestinal tract."

Sarcoma is derived from the Greek *sark*, meaning "flesh," and *oma*, meaning "tumor." These growths originate in the same kinds of cells that give rise to connective tissue, muscle, and bone. In spite of Kaposi's admonition about his disease that "the prognosis is unfavorable . . . and fatal termination could not be prevented by extirpation, local or general, or the administration of arsenic [a favored treatment for cancer at the time]," physicians for a century underestimated the danger of this unusual malignancy.

Because the progression of KS was known to be slow, requiring "three to eight years or more," subsequent textbooks most commonly employed the word *indolent* to describe its course. Thus was conveyed an erroneous message about the basically lethal nature of the malignancy, even though some authorities continued to write of its deadly manifestations, such as massive intestinal bleeding. The word *indolent*, in fact, appears in the original 1981 reports in British and American medical journals of outbreaks of Kaposi's sarcoma among gay men. So alarmed, however, were the authors of those reports by the sudden raging aggressiveness of a disease traditionally regarded as lethargic that the American article saw fit to remind its readers that the course had sometimes been known to be "fulminant, with extensive visceral involvement"; the paper published in England made the same case and gave it immediacy by pointing out that "half our patients were dead within 20 months of diagnosis." Clearly, this was a new form of KS, suddenly far more worrisome than even Kaposi had warned.

Decades before KS became associated in physicians' minds with HIV infection, it was being recognized too often for coincidence as an accompaniment of various forms of the lymphatic cancer called lymphoma. Today, KS and lymphoma, not necessarily concomitant, are the two leading malignancies that prey on people with AIDS. Except for reasons of immunodeficiency, the relationship between the two has not yet been clarified. AIDS-related lymphoma, which most often involves the central nervous system, gastrointestinal tract, liver, and bone marrow, is no less aggressive than KS.

Unlike any other pestilence previously known to humankind, HIV does not limit its array of deadly options. There are only so many ways for a pancreatic cancer, for example, to kill; when a heart fails, or a kidney, very specific events take place; a deadly stroke takes aim at a single focus in the brain, starting its victim down a well-marked road to deterioration. Not so with HIV—it offers seemingly endless choices as one organ system and then another is set upon by a wide assortment of microbes and cancers. At autopsy, the only consistently predictable finding is a severe depletion of the lymphatic tissue that is part of the immune system. At the dissecting table, even members of the AIDS care team are often surprised by unexpected areas of involvement and the degree to which the tissues of their patient have been laid waste.

Respiratory failure, sepsis, destruction of brain tissue by tumor or infection—these are the more common immediate causes of death; some patients bleed into the brain, or the lung, or even the gastrointestinal tract, and some succumb to widespread tuberculosis or sarcoma; organs fail, tissues bleed, infection is everywhere. And invariably, there is malnutrition. No matter the magnitude of methods activated to fight it, starvation cannot be prevented. A care unit for terminal AIDS patients is peopled by emaciated, wraithlike men and women whose shrunken eyes look dully out from cavernous sockets, their faces often without expression, their bodies wizened with the shriveled frailty of prematurely advanced age. Most are beyond courage. The virus has robbed them of their youth, and it is about to rob them of the rest of their lives.

Autopsy pathologists distinguish between two separate designations for the cause of death: They refer to the proximate cause of death and the immediate cause of death (officially known by the acronyms PCOD and ICOD). For all of these young people, the PCOD will be AIDS—the specific ICOD seems hardly to matter. The quantity of suffering is the same for all, even though the quality varies. I talked about these matters not long ago with Dr. Peter Selwyn, one of the several Yale professors whose single-minded devotion to the care of AIDS patients has animated the efforts of many of the residents and students at our school. Despite his authoritative contributions to current understanding of HIV infection, he is a reticent man who expresses large concepts with few words. He said simply, "My patients die, I think, when their time comes." It seemed an incongruous statement, floating there in the company of the biomedical complexities still hanging in the air from our long discussion of molecular biology and bedside management. And yet it made sense. At the end, he said, so many things go wrong that there comes a time when the depleted forces of gasping life just seem to give out. Death comes with sepsis and organ failure and starvation and with the final departure of the spirit, all at once. Selwyn has seen it many times, and he knows.

I am a hundred miles away from the hospital. This is one of those unexpected afternoons in mid-autumn when everything under nature's cloudless blue sky has become exactly the way it should be, but almost never is. The summer just passed was rainy, and perhaps for that reason the hills surrounding my friend's farm have taken on those heartbursting effusions of color that are almost more than my city-bred soul can comprehend or contain. Nature is being kind without knowing it, as nature can be cruel without knowing it. At such an instant, it seems as though no other day will ever attain the impossible splendor of this one. Already, I feel a nostalgia for today even as I live it. I am obsessed with an urge to memorize the image of every tree because I know its blazing flourish will begin to fade as soon as tomorrow, and never appear precisely like this again. When a thing is beautiful and good, it should be seen so clearly, and held so snugly, that no one will ever forget how it looks and how it feels.

I am sitting in the sunny kitchen of John Seidman's farmhouse, built a century ago in the midst of twenty acres of fertile land, near the town of Lomontville in upper New York State. In an upstairs bedroom, ten years ago, John's best friend, David Rounds, died in his arms at the end of a long and difficult illness. John and David were more than best friends; they shared a love that was meant to endure. But cancer determined otherwise. David was taken from John, and from those others of us who also loved him in our variety of ways, at a time when the future seemed secure and certain for both of them. David had won a Tony Award for Best Supporting Actor on Broadway only two years before, and John's stage career was showing increasing promise. In that farmhouse, grief was a long time in passing before life resumed its proper rhythms.

I have known John Seidman for almost twenty years, and Sarah, my wife, shared a house with him and David long before that. He has been so close a friend to my family that my two youngest children call him Uncle. And yet there is a large part of his life that he and I have never discussed and about which I know almost nothing. On this splendorous day just before the fleeting grandeur of autumn disappears, the two of us are sitting together and we are talking about death—and AIDS.

Death has become much too familiar to John. It is as though the loss of David was the prelude to a succession of sorrows, during which friends, colleagues in the theater, and even mere acquaintances sickened, withered, and died. In the past decade, John has repeated with one after another the cycle of discovery of seropositivity, disease progression, watchful caregiving, descent to terminal illness, and death—again and again. In his early forties, he is one of tragedy's witnesses. There have been many others, and more than a few are now dead. The young men, and the few young women, who have companioned one another to the grave have been taken in the most productive years of their lives—what might have been and what should have been is lost. The vigor, the talent, and undoubtedly the genius of a generation are diminished, and so is our society.

We talk about John's friend Kent Griswold, who died in 1990

with toxoplasmosis and a trio of the common acronyms: CMV, MAI, and several bouts of PCP. Could there, I wanted to know, be any dignity in such a death? Can anything be salvaged of what once was, to bring a sense of himself to a man near his final hour, when he has been through so much? John thought a long time before answering, not because he had never considered the question before but because he wanted to be sure I would understand. The search for the elusive dignity, he said, may become irrelevant to the person who is dying—he has already carried out his struggle, and so often near the end, those around him can detect no recognizable conscious thought. Dignity is something, said John, that the survivors snatch—it is in their minds that it exists, if it exists at all:

> Those of us left behind search for dignity in order not to think ill of ourselves. We try to atone for our dying friend's inability to achieve a measure of dignity, perhaps by forcing it on him. It's our one possible victory over the awful process of this kind of death. With a disease like AIDS, we need to deal with the sadness that comes with seeing a beloved friend lose his particularity, his uniqueness. Toward the end, he becomes just like the last person you saw go through this. You feel the sadness of seeing someone lose his individuality and become a clinical model.
>
> How much of the "good death" is for the person dying and how much for the person helping him? They're obviously related to one another, but the question is how. To me, the concept of a good death is generally not something that can be managed very well for the one who is dying. A "good death" is only a relative thing, and what it really means is decreasing the mess. There isn't much you can manage beyond trying to keep things neat and keep things painless—keeping someone from being alone. But leading up to those final moments, I think even the importance of one's not being alone is something we infer.
>
> In retrospect, and in a way this sounds brutal, my own experience is that the only means we have of knowing if we have helped someone to a better death is whether or not we feel regret, or whether there is anything we feel sorry about or have left undone. If we can truly say that we missed no opportunity to do what we could, we've done the best job that's possible. But

even that, as an absolute achievement, only has absolute value to oneself. What you're left with at the end is a situation that makes no one happy. The fact is that you've lost someone. There's no way to feel good about it.

The one bond we do need to believe is absolutely unbreakable in death is love. If love is what we feel we're providing at those mysterious moments leading up to death, that, I suppose, is what makes a death "good," if anything really can. But it's such a subjective quality.

During his terminal weeks in the hospital, Kent was never alone. Whatever help they could or could not provide him at the final hours, there is no question that the constant presence of his friends eased him beyond what might have been achieved by the nursing staff, no matter the attentiveness of their care. It is impossible to observe homosexual AIDS patients without being struck by the way, a circle of friends, not necessarily all gay, will almost predictably come together as a man's family and assume responsibility for what a wife or parents might otherwise do. Dr. Alvin Novick, one of the earliest of America's AIDS activists and among the most highly respected, has called this phenomenon of joined commitment "the caregiving surround." It is a communal act of love, but it is also something more. John describes it:

AIDS is happening to people, especially in the case of gay men, who have created families by a conscious affinity—we have chosen the people who will be our family. Our sense of responsibility to each other isn't based on the usual social forms. In many cases, the traditional family has rejected us. So the affinitive family is much more important.

A lot of the greater community really do feel that what is happening to us *should* happen to us—it's some kind of visitation on us for our sinful and abnormal ways. And so it's in our mutual interest not to leave someone alone with that judgment of society. Those of us who suffer from some kind of self-loathing may find it very easy to think of AIDS as a form of punishment, but even those of us who don't are aware that much of society does see it that way. To neglect our friends who have to deal with the disease themselves is somehow to abandon them to the judgment of the straight world.

Kent's last few weeks, John tells me, were like those of so many other people with AIDS, and of so many people with any of the diseases that slowly eat away at life's ebbing strength. Following the long months during which he had been forced to fight off one unanticipated problem after another, he seemed to undergo a suspension of any understanding that an incremental reduction of control was accompanying each new complication. As he stopped trying to comprehend, he also stopped struggling against the successive assaults, as though it seemed now less important to resist—there was no longer any point to it. Or perhaps the effort required to grasp the significance of events simply sapped too much of his limited energy.

The details of a latest onslaught lost their urgency. There are those who would call such exhausted indifference acceptance, but the very word implies a welcoming. Perhaps it is rather the recognition of defeat, the involuntary acknowledgment that the time has come to quit fighting. Most of the dying, not only of AIDS but of any prolonged sickness, seem unaware that they have reached this stage. For some few, mental faculties remain so intact that they are able to consciously decide, but much more often the decision is made for them by a lapse into a degree of lessened sensibility or even coma. This is the phase of dying in which William Osler and Lewis Thomas seldom saw aught but serenity. For most of us, it will come much too late to give consolation to those who watch at the bedside.

While Kent had been less sick, he had sometimes spoken of his concern about how much physical pain he would be able to withstand, how uncomfortable his last weeks might be. He expressed a wish to find that critical moment when he could knowingly make up his own mind whether to continue the struggle. No one around him could tell for sure whether that wish was granted.

An influential friend had somehow gotten Kent a commodious private room in the hospital, and in that large space he appeared daily to become smaller. He seemed almost hard to find. In John's words, "He dwindled further and further beneath the sheets." Even when he was at his strongest, Kent needed help to get to the bathroom, but the rest of the time he was completely bedridden. Never a large man, he seemed now to be disappearing. As John describes Kent's withering, I think again of Thomas Browne watching his dying friend go through the same process 350 years earlier: "He came to be almost half himself and left a great part behind him which he carried not to the grave."

Because of Kent's toxoplasmosis, he was losing cognition to the point where he was unable to comprehend what was going on around him. CMV retinitis blinded first one eye and then the other. He had by then wasted to such a degree that it was impossible to read his face or decipher his expressions—was he smiling, or was it a grimace that twisted the corners of his silent mouth? John says it so well: "A form of communication is lost when someone is so diminished." The dying man's whole body had grown very dark, especially his face.

Early on, Kent had made it clear that no aggressive treatment was to be used once it became evident that it would be futile. Guided by that, his "caregiving surround" consulted with the doctors, and together they tried to make correct decisions as each succeeding necessity arose. Finally, there were no longer any decisions to make. It had become so clear—there was nothing further to be done. It was just as Peter Selwyn says—Kent's time had come.

Kent was less and less conscious of any sort of discomfort. No longer was it important that he receive medical help of any kind. "It became our mission just to keep him surrounded, just to keep him connected, at least as much as he was capable of sensing any connection. The most important thing was that we didn't want him to be alone." At the end, Kent just slipped away. John now comes to the final part of the story.

> I wasn't in New York when he died—I was up here at the farm for a few days. I got off a bus at the Port Authority and called in to my machine. There was a message that Kent was gone, and it shocked me. When I'd last seen him, he wasn't recognizably living, certainly not recognizably Kent. Even though he was expected to die any minute, somehow the idea that he was actually gone—I suppose the shock had to do with the fact that after all the time I had spent with him I had to find out about it in that rotten way—standing there alone in a grimy phone booth, hearing it from my answering machine.

Kent died among companions who had helped sustain him in his last two years of life. He was not one of the many homosexuals and drug abusers who have been ostracized by their families—he was the only child of older parents and they had died years before. Without the devotion of his friends, his death, and his life as well, would have been soon forgotten.

Nothing that has been written here should be construed to imply that traditional families are only seldom involved in the care of their sons and daughters (or husbands and wives) dying of AIDS. Quite the opposite is true. Gerald Friedland describes the return, the reuniting of parents, mothers in particular, with the children whose lives and friends they had years before rejected. This is true not only of the families of gay men but of those of drug abusers as well. Of course, not all homosexuals and not all drug abusers have separated from their roots in the first place, and so it is not uncommon that the last months of a young man or woman with AIDS are spent in the nurturing care of siblings or parents, sometimes together with a small group of their child's friends or a lover. It is usually a great deal easier for a middle-class parent to leave employment or a distant home than it would be for members of a family from an inner-city ghetto or barrio, where even a day's absence from work means loss not only of income but possibly of the low-paying job itself. I have been told of mothers with as many as four children at once dying of AIDS—the cruelty of the virus reaches magnitudes beyond any imaginable reality.

At the bedsides of dying young people watch mothers and wives, husbands and lovers—sisters, brothers, and friends—doing what they can to buffer the onslaughts of messy death. As in ages past when a child is mortally ill, murmuring voices of parents are heard, sometimes barely audible in the hush that precedes a life's departure. They are soft words of encouragement and they are prayers. In English or in Spanish, and in other languages throughout the world, there has been repeated so many times one variant or another of the words spoken by the biblical King David as he wept over the body of his slain son, the rebellious Absalom, from whom he had been for so many years estranged:

> O my son Absalom,
> my son, my son Absalom!
> Would God I had died for thee,
> Oh Absalom, my son, my son.

Gerald Friedland speaks of the "inversion of the expected life cycle"—parents are burying their children. An aberration has recurred from earlier centuries, just when we had complacently concluded that our science had conquered it. Not only the virus is turned back to front but so is the pattern of natural logic by which the young should bury the old. There is finally a metaphoric lesson here—in the therapy that is at present our best means of inhibiting the propagation of HIV: With AZT and other drugs, we try to stop the reverse transcriptase, and thereby stop the reversal, too, that turns the cycle of life on its head. Our scheme works, but not as well as we would like, and death continues to pursue the young and even the very young, while their elders can only stand by and mourn.

What dignity or meaning can be snatched from such a death will never be known, except by those whose lives have embraced the life just lost. The young people who provide the hospital care for the young people who die—and here I refer not only to doctors and nurses but to every one of those dedicated personnel—look on and wonder that such selflessness exists in a world they have been taught is cynical. Their own daily deeds belie the cynicism—they, too, are heroes of a sort. Their heroism is contemporary and unique to the path they have chosen as health workers who conquer their own fears and vanquish their feelings of vulnerability for the sake of those afflicted with AIDS. They make no moral judgments—they do not distinguish between social classes, manners of infection, or memberships in those categories called risk groups. Camus described it well: "What's true of all the evils in the world is true of plague as well. It helps men to rise above themselves."

In the midst of the stories that still come to us of unwilling physicians here or HIV-phobic surgeons there (and the more than 20 percent of surveyed American medical residents who would treat people with HIV but if given the opportunity would choose not to), it is heartening to know that the AIDS-affected are being watched over by people like these. For our children who care for our children stricken by HIV, the burden is made still greater with the sorrow of being stewards to the mortality of men and women their own age or perhaps only a decade older. In that injustice lies the source of the most outraged of those many reproaches we hurl at insensate nature, whose mindless tinkering created HIV—that it robs us of great pieces of the fabric from which we are entitled to fashion our future. Of the youthful legions lost to AIDS, it is proper to speak the words written seventy years ago by the neurosurgeon Harvey Cushing as he grieved over his companions martyred in World War I. They are, Cushing lamented, "doubly dead in that they died so young."

X

The Malevolence of Cancer

> Once upon a time, there was a little chimney-sweep, and his name was Tom. That is a short name, and you have heard it before, so you will not have much trouble in remembering it. He lived in a great town in the North country, where there were plenty of chimneys to sweep, and plenty of money for Tom to earn and his master to spend. He could not read nor write, and he did not care to do either; and he never washed himself, for there was no water up the court where he lived. He had never been taught to say his prayers. He never had heard of God, or Christ, except in words which you have never heard, and which it would have been well if he had never heard. He cried half his time and laughed the other half. He cried when he had to climb the dark flues, rubbing his poor knees and elbows raw; and when the soot got into his eyes, which it did every day in the week; and when he had not enough to eat, which happened every day in the week likewise.

So begins Charles Kingsley's 1863 children's classic, *The Water Babies.* Tom was what the English gentry euphemistically called a "climbing boy." His duties required no lengthy training and there were no prerequisites for entering the profession. Most recruits took up the depressing occupation between the ages of four and ten. Each day's work was launched simply enough: "after a whimper or two, and a kick from his master, into the grate Tom went, and up the chimney."

Those chimneys bore little resemblance to the straight uprights of a later architectural style. Even by Kingsley's day, the mid-

1800s, they had become more direct in their ascent than they had been when the British surgeon Percivall Pott turned his attention to their dangers in 1775. In Pott's time they not only were tortuous and irregular but had an annoying habit of running in a horizontal direction for short distances before resuming their intended vertical course. The result of all the structural peregrinations was that there were plenty of nooks, crannies, and flat surfaces upon which soot would accumulate. Not only that, but a climbing boy's squirming en route up the flue made it quite likely that he would abrade the skin surfaces on various parts of his body, especially those that projected or hung.

The word *hung* is used deliberately here to mean exactly what it sounds like; more often than not, the little climbers did their grimy work without the protection of any layer of clothing between themselves and the filthy walls along which they scrambled. They were quite naked. There was a good and sound tricks-of-the-trade reason for the vocational nudity, or at least the boys' masters thought it was good and sound. The chimneys were very narrow, measuring approximately twelve to twenty-four inches in diameter. Why go to all the trouble of finding such small, skinny lads if they were only going to use up valuable space by wearing clothes? So the master sweep recruited the tiniest boys he could find, taught them the rudiments of chimney-shinnying, and kicked their bare, coal-blackened bottoms into the gratings each morning, shouting them up the tight, airless shafts to start the day's work.

The problems were compounded by the personal habits of the poor sweeps themselves. Coming as they did from the very lowest stratum of the English social structure, they had never learned to value bodily cleanliness. Moreover, many of these unfortunate lads, in spite of being exposed to such a great deal of hearth, had never known much home. There had been no loving maternal hands to guide them, or even to pull them by the ears to a warm tub. By and large, they were abandoned urchins. The tar-laden particles remained buried in the wrinkles and folds of their scrotal skin for months at a time, relentlessly eating away at their lives while the cruelties of their masters ate away at their souls.

Percivall Pott (1714–1788) was the most distinguished London surgeon of his generation, and he knew a great deal about the difficult life of the young English sweeps. He observed that "The fate of these people seems singularly hard: in their early infancy, they are most frequently treated with great brutality, and almost starved with cold and hunger; they are thrust up narrow, and sometimes hot chimnies, where they are bruised, burned, and almost suffocated; and when they get to puberty, become peculiarly liable to a most noisome, painful, and fatal disease." These words were written in 1775; they appeared in a brief section of a much longer article by Pott, entitled "Chirurgical observations relative to the cataract, the polypus of the nose, the cancer of the scrotum, the different kinds of ruptures and the modification of the toes and feet." This article contains the first description ever recorded of an occupational malignancy. The disease took years to develop, but it sometimes began to make its appearance as early as the time of puberty. In the first decade of the nineteenth century, it was reported in a child of eight.

There is no doubt that Pott was describing a fatal malignancy that we would nowadays call squamous cell carcinoma. What he observed in the scrotums of his young patients was "a superficial, painful, ragged, ill-looking sore, with hard and rising edges: the trade call it the soot-wart. . . . It makes its way up the spermatic process into the abdomen. . . . When arrived within the abdomen, it affects some of the viscera, and then very soon becomes painfully destructive."

Pott well knew that the scrotal cancer, except in the few cases when it was surgically excised at a very early stage, killed every one of its victims. He had attempted surgical cure over and over again, even though it meant, in those horrible days before the invention of anesthesia, that he had to strap a screaming boy down onto a table and keep him immobilized in the grip of powerful assistants. Eligibility for the operation was restricted to those boys in whom the ulcerating process was still limited to one side.

The procedure perpetrated as great an assault on the psyche as it did on the soma, consisting as it did of slashing away as rapidly as possible the testicle and half the scrotum of these unfortunate adolescents. Bleeding tissues were treated by pressing a red-hot iron directly into them. Attempting to stitch the hideous charred wounds always caused a pus-laden infection, so the surgical area was left open to drain sloughed detritus and fluid throughout the long months of slow healing.

Pott's results did not often justify the ordeal. He was disheartened by long-term follow-up studies of his patients: "But though the sores, after such operation, have in some instances healed kindly, and the patients have gone from the hospital seemingly well, yet, in the space of a few months, it has generally happened, that they have returned either with some disease in the other testicle, or in the glands of the groin, or with such wan complexions, such a total loss of strength, and such frequent and acute internal pains, as have sufficiently proved a diseased state of some of the viscera, and which have soon been followed by a painful death." Although Pott's commas may be extravagant, his description is not. If anything, he understates the miseries with which these boys went to the grave.

Pott recognized that this dreadful courier of death began as an abnormal growth restricted to a specific location, a process which only later began that relentless creeping course of foul ulceration by which it infiltrated its rotting pathway into the structures around it. He published his case studies at a time favorable to the formulation of a thesis that involved the influence of intrusive foreign materials on the body. A few prominent medical theorists had recently begun to introduce the concept that living tissue requires a stimulus, which they called an "irritation," in order to cause it to perform its normal functions. From the principle of irritation, it was only a short step to the concept that diseased organs are sick because a part or all of them have become inflamed—in other words, overirritated. Pott argued that cancer in the private parts of chimney sweeps was the direct result of inflammation caused by the chemical action of soot.

These days, the Surgeon General's warning stares out at us from every cigarette ad, and there are not many people who don't take it seriously. No literate American adult is unaware of the cancer-causing properties of tars and resins, and most understand that those properties arise from the chemical irritation produced in living tissue by constant contact with the noxious substances. But as self-evident as it seems today, the concept that chronic irritation may cause disease was not always appreciated by physicians. At the time when Percivall Pott chose to go beyond mere clinical description of scrotal cancer by stating his conviction that it was the result of a highly specific response to soot, the theory of irritation and inflammation was still on very shaky ground, and indeed most of it was later discarded. Although the sweeps themselves called their malady the "soot-wart," they seem not to have grasped the idea that cleansing themselves of the grime with an occasional wash might prevent it. They simply accepted as inevitable that a certain number of them would develop the condition and suffer an agonizing death—the risk came with the job.

Pott's thesis that soot was the instigating cause of the cancer received immediate recognition. It led directly to a parliamentary decree that no chimney sweep might start his apprenticeship before he was eight years old and that all of the boys must be given a bath at least once a week. By 1842, no boy under the age of twenty-one was permitted to climb chimneys. Unfortunately, the law was so often broken that there were still plenty of under-age sweeps when Charles Kingsley was writing *The Water Babies* twenty years later.

As early as the days of Hippocrates and even before, the ancient Greek physicians had a clear understanding of the ways in which a malignant growth so often pursues its inexorable determination to destroy life. They gave a very specific name to the hard swellings and ulcerations they so commonly saw in the breast or protruding from the rectum or vagina; they based that name on the evidence of their eyes and fingers. To distinguish them from ordinary swellings, which they called *oncos*, they used the term *karkinos*, or "crab," derived, interestingly enough, from an Indo-European root meaning "hard." *Oma* being a suffix referring to "tumor," *karkinoma* was used to designate a tumorous growth that was malignant. Centuries later, the Latin word for "crab," *cancer*, came into common usage. *Oncos*, meantime, came to be applied to tumors of any kind, which is why we call a cancer specialist an oncologist.

Karkinoma was said to be due to the stagnation within the body of an excess of a hypothetical fluid called black bile, or *melan cholos* (from *melas*, "black" and *chole*, "bile"). Since the Greeks did not dissect the human body, the cancers they saw were ulcerated malignancies of the breast or skin, and those of the rectum and female genital tract which had grown so large that they protruded through body openings. Accordingly, the fanciful explanation was supported by the common observation that cancer patients were indeed melancholy, and for obvious good reason.

The origin of *karkinos* and *karkinoma* was based, as were so many Greek medical terms, on simple observation and touch. As Galen, the foremost interpreter and codifier of Greek medicine, put it in the second century A.D., the appearance of this creeping, infiltrating stony mass, ulcerated at its center, which he so often saw in the breasts of women, is "just like a crab's legs extending outward from every part of its body." And it is not only the legs that are digging farther and deeper into the flesh of its victim—the center, too, is eroding its way directly through her.

The likeness is to an insidious, groping parasite, attached by sharp-clawed tentacles to the decaying surface of its imperiled prey. The clawing extremities ceaselessly extend the periphery of their malign grip, while the loathsome core of the burrowing beast eats silently away at life, able to digest only what it has first decomposed. The process is noiseless; it has no recognizable instant of beginning and it ends only when the despoiler has consumed the final remnants of its host's vital forces.

Until after the middle of the nineteenth century, cancer was thought to do its killing by stealth. Its lurking force lay under the cover of hushed darkness, its first sting felt only when murderous infiltration had strangled too much normal tissue to restore the overwhelmed defenses of its host. The perpetrator regurgitated as malignant gangrene the life it had noiselessly chewed up.

We know better now, because we have come to recognize a different personality when our old enemy is seen through the microscope of contemporary science. Cancer, far from being a clandestine foe, is in fact berserk with the malicious exuberance of killing. The disease pursues a continuous, uninhibited, circumferential, barn-burning expedition of destructiveness, in which it heeds no rules, follows no commands, and explodes all resistance in a homicidal riot of devastation. Its cells behave like the members of a barbarian horde run amok—leaderless and undirected, but with a single-minded purpose: to plunder everything within reach. This is what medical scientists mean when they use the word *autonomy*. The form and rate of multiplication of the murderous cells violate every rule of decorum within the living animal whose vital nutrients nourish it only to be destroyed by this enlarging atrocity that has sprung newborn from its own protoplasm. In this sense, cancer is not a parasite. Galen was wrong to call it *praeter naturam*, "outside of nature." Its first cells are the bastard offspring of unsuspecting parents who ultimately reject them because they are ugly, deformed, and unruly. In the community of living tissues, the uncontrolled mob of misfits that is cancer behaves like a gang of perpetually wilding adolescents. They are the juvenile delinquents of cellular society.

Cancer is best viewed as a disease of altered maturation; it is the result of a multistage process of growth and development having gone awry. Under ordinary conditions, normal cells are constantly being replenished as they die, not only by the reproduction of their younger survivors but also by an actively reproducing group of progenitors called stem cells. Stem cells are very immature forms with enormous potential to create new tissue. In order for the progeny of the stem cells to progress to normal maturity, they must pass through a series of steps. As they get closer to full maturity, they lose their ability to proliferate rapidly in proportion to the increase in their ability to perform the functions for which they are intended as grown-ups. A fully mature cell of the intestinal lining, for example, absorbs nutriments from the cavity of the gut a lot more efficiently than it reproduces; a fully mature thyroid cell is at its best when it secretes hormone, but it is much less inclined to reproduce than it was while younger. The analogy with the social behavior of a whole organism, like us, is inescapable.

A tumor cell is one that has somewhere along the way been stopped in its capacity to *differentiate*, which is the term used by scientists for the process by which cells go through the steps that enable them to reach healthy adulthood. The clump of immature abnormal cells that results from the blocking of differentiation is called a neoplasm, derived from the Greek word for a new growth or formation. In modern times, the word neoplasm is used synonymously with *tumor*. Those tumors whose cells have been blocked closest to the attainment of the mature state are the least dangerous and are therefore called benign. A benign tumor has retained relatively little of its potential for uncontrolled reproduction—it is well differentiated; under the microscope, it looks a lot like the adult it was close to becoming. It grows slowly, does not invade surrounding tissues or travel to other parts of the body, is often surrounded by a distinct fibrous capsule, and almost never has the capacity to kill its host.

A malignant neoplasm—what we call cancer—is a different creature entirely. Some influence or combination of influences, whether genetic, environmental, or otherwise, has acted as the triggering mechanism to interfere so early in the pathway of maturation that the progress of the cells has been stopped at a stage when they still have an infinite capacity to reproduce. Normal stem cells keep trying to produce normal offspring, but their development continues to be arrested. They do not attain a sufficient level of adulthood to do the work they were meant for or to look more than just a little like the grown-up forms they were intended to be. Cancer cells are fixed at an age where they are still too young to have learned the rules of the society in which they live. As with so many immature individuals of all living kinds, everything they do is excessive and uncoordinated with the needs or constraints of their neighbors.

Being not completely grown-up, a cancer cell does not engage in some of the more complicated metabolic activities of mature nonmalignant tissue. A cancer cell of the intestine, for example, doesn't help out in digestion as its adult counterpart does; a cancer cell of the lung is uninvolved in the process of respiration; the same is true of almost all other malignancies. Malignant cells concentrate their energies on reproduction rather than in partaking in the missions a tissue must carry out in order for the life of the organism to go on. The bastard offspring of their hyperactive (albeit asexual) "fornicating" are without the resources to do anything but cause trouble and burden the hardworking community around them. Like their progenitors, they are reproductive but not productive. As individuals, they victimize a sedate, conforming society.

Cancer cells do not even have the decency to die when they should. All nature recognizes that death is the final step in the process of normal maturation. Malignant cells don't reach that point—their longevity is not finite. What is true of Dr. Hayflick's fibroblasts does not apply to the cellular population of a malignant growth. Cancer cells cultivated in the laboratory exhibit an unlimited capacity to grow and generate new tumors. In the words of my research colleagues, they are "immortalized." The combination of delayed death and uncontrolled birth are malignancy's greatest violations of the natural order of things. These two factors in combination are the main reasons a cancer, unlike normal tissue, continues to enlarge throughout its lifetime.

Knowing no rules, cancer is amoral. Knowing no purpose other than to destroy life, cancer is *im*moral. A cluster of malignant cells is a disorganized autonomous mob of maladjusted adolescents, raging against the society from which it sprang. It is a street gang intent on mayhem. If we cannot help its members grow up, anything we can do to arrest them, remove them from our midst, or induce their demise—anything that accomplishes one of those aims—is praiseworthy.

There comes a point at which home turf is not enough—offshoots of the gang take wing, invade other communities, and, emboldened by their unresisted depredations, wreak havoc on the entire commonwealth of the body. But in the end, there is no victory for cancer. When it kills its victim, it kills itself. A cancer is born with a death wish.

Cancer is, in every possible sense, a nonconformist. But, unlike some nonconformist individuals about whom there is much to admire, the nonconforming malignant cell has not a single redeeming feature. It does everything it can not only to disassociate itself from but even to destroy the community of cells that has given it life. As though to make certain that it is not confused with the conformist adult members of its original family, the cancer cell retains an immature and different appearance and even shape. This characteristic of malignant growth is called anaplasia, from the Greek term meaning "without form." The anaplastic cell gives birth to anaplastic offspring.

But try as it may, only an unusual cancer is composed of cells that have changed their appearance completely enough to become unrecognizable as members of their own original tribe. Except in extreme cases, a careful look down the barrel of a microscope at a bit of the diseased tissue will suffice to reveal its ancestral lineage. Thus, a bowel cancer can be identified as what it is because it still has some characteristic features that betray its intestinal origin. Even far away from home, as when the bloodstream has carried its cells to the liver, the cancer's face, almost no matter the degree of anaplasia, will usually give it away. Even cancer, that remorseless renegade that ran away to join the biological equivalent of Murder, Inc., retains some dimly recognizable traits of its old family and its old obligations.

The twin characteristics of autonomy and anaplasia define the modern understanding of cancer. Whether they are to be thought of as "ugly, deformed, and unruly" or more academically as "anaplastic" and "autonomous," the cells of a cancer are wicked in ways far beyond what is implied by the scientific connotation of the word *malignant*. *Malevolent*, in fact, says it better, because it bears the implication of an element of ill will.

The deformity and ugliness of the individual cancer cell are most manifest in the irregularities of its distorted shape. Whereas the appearance of a normal cell in normal tissue differs hardly at all from that of its normal neighbors, the forms and dimensions of the individuals in a cancer's cellular population are usually neither uniform nor orderly. They may bulge, flatten, elongate, round themselves out, or in some other way demonstrate that each is created as though with a mind of its own—it is an independent agent. Cancer is a state in which a breakdown has occurred in the communication and mutual interdependence between cells. That sequence of events noted above has taken place, in which the genetic characteristics of the malignant cell become altered, and everything else about the disease follows from that fact. Some of the environmental, lifestyle, and other causes of the alterations are known, some are being studied, and some are no doubt still unsuspected.

Though chaotic in appearance and inconsistent in size, the community of malignant cells is not necessarily always anarchic. In a few forms of cancer, in fact, all individuals are found to choose a specific uniform shape that suits a shared element in their willfulness. Such malignancies exist as though to demonstrate an obstinate refusal to conform to the accustomed disharmony expected of them; their cells reproduce myriads of virtually identical selves, like so many millions upon millions of little poisonous apples, boringly similar to one another but quite different from their tissue of origin. Even the predictability of malignancy's unpredictability is unpredictable.

The central structure of the cancer cell, its nucleus, is larger and more prominent than that of mature relatives and is often as misshapen as the cell itself. Its dominance over the protoplasm surrounding it is intensified by the enhanced avidity with which it takes up standard laboratory stains, a characteristic that gives it a darkened, ominous look. The evil-eye nucleus reveals its disordered independence in yet another way: Instead of dividing neatly into two symmetrical halves during the process of reproduction known as mitosis, the chromosomes (the components of the nucleus that carry the DNA) align themselves in bizarre patterns, attempting with varying degrees of success to multiply, figuratively head over heels, without any element of precision or accountability. The rate of mitosis of some cancers is so rapid that a quick look through the microscope will catch many times the number of cells in the act of trying to reproduce as are found in mature normal tissue, and every one of them seems to be doing it in its own haphazard way. Small wonder that the surviving offspring are ill-suited to their surroundings in the ordered, consistent tissue of the organs of which they were originally meant to be a part. So pugnaciously "other" are the new masses of cells, in fact, that they not only invade but also push their law-abiding grown-up neighbors out of the way as they infiltrate and preempt surrounding territory.

In a word, cancer is asocial. Having escaped the constraints that govern nonmalignant cells, the newly formed tissues pursue uncontrolled and domineering relationships with their host organs and cannot be made to restrict their encroaching margins to the foci that gave them birth. Unrestrained and patternless growth enables a cancer to force its way into nearby vital structures to engulf them, prevent their functioning, and choke off their vitality. By this means, and by destroying the organs from whose stem cells they are made, the masses of cancer cells kill the gradually sickening person after feasting on the nutrients that were to have sustained him.

Although it begins as a microscopic phenomenon, the process of malignant growth, once properly established, inevitably continues until it can be seen with the naked eye or felt with the exploring hand. For a while, the growing mass may remain too small or confined to produce symptoms, but in time, the cancer's victim will sense that something untoward is happening to him. By that point, the malignancy may have grown so large that it is beyond cure. Particularly in certain solid organs, a cancer may reach considerable size before it makes its host aware of its presence. It was for this reason, of course, that the disease achieved its legendary reputation as a noiseless killer.

A kidney, for example, may be found to harbor a perfectly huge growth when it first reveals its advanced state of disease by spilling visible blood into the urine or causing a dull ache in the flank. If an operation is done at that point, the surgeon's efforts will be defeated by the wide extent of involvement of surrounding tissues. The otherwise-symmetrical brown smoothness of the organ will be found to have been eaten away in one large area by an ugly, lobulated protrusion of coarse gray hardness that has forced its way through to the surface, invaded the adjacent fat, and drawn all nearby tissues into it, the misbegotten whole forming one great puckered grotesquerie of bunched-up aggression. Of all the diseases they treat, cancer is the one that surgeons have given the specific designation of "The Enemy."

The visible structure and invasiveness of a cancer are only two of its many forms of unruliness. One of the most duplicitous of malignancy's misbehaviors is the way in which it seems to elude the defenses ordinarily mounted by the body against tissue it perceives as not belonging to it. Theoretically at least, cells that have become cancerous should be detectable as foreign or "other" by an intact immune system and then killed, much as is a virus. This actually does happen to an extent; some researchers believe that our tissues are continually making cancers, which are just as continually being destroyed by this kind of mechanism. Clinical malignancies would then develop in those rare instances when the surveillance system fails. An example of support for such a thesis is to be found in the prevalence in people with AIDS of tumors such as lymphoma and Kaposi's sarcoma. Overall, the incidence of malignancies in immunocompromised individuals is some two hundred times that found in the general population, and for Kaposi's the figure is more than twice what it is for the average tumor. One of the most promising fields of today's biomedical research is the study of tumor immunity with a view toward strengthening the body's responses to the antigens that cancers may produce. Although there have been some promising results, the target cells continue, for the most part, to outwit the scientists.

Normal cells require a complex mixture of nutrients and growth factors in order to continue functioning and retain viability. Throughout all tissues of the body, they are bathed in a life-giving nutrient soup called extracellular fluid, which is constantly being restored and cleansed by exchanging substances with circulating blood. The blood's plasma, in fact, amounts to one-fifth of the body's extracellular fluid; most of the other four-fifths lies between the cells, and is called interstitial. The interstitial fluid accounts for approximately 15 percent of body weight; if you weigh 150 pounds, your tissues are soaking in 22 pints of the salty stuff. The nineteenth-century French physiologist Claude Bernard introduced the term *milieu intérieur* to name and describe the function of this internal environment in which cells live within us. It is as though the earliest groups of prehistoric cells, when they first began to form complex organisms in the marine depths from which they drew sustenance, brought some of the sea into and around themselves so that they might continue to be nourished by it. Among the unique features of malignant tissues is their reduced dependence on the nutritional and growth factors in the extracellular fluid. Their lessened need for sustenance from the surroundings enables them to grow and invade even those areas beyond optimal supply lines.

No matter that each cellular unit can get along with less, the helter-skelter increase in population soon accumulates so many malignant cells that the requirements of the aggregate tend to outstrip whatever supplies are available. As a result, a total tumor mass will often develop an increased demand for nutrition, even though each individual within it may require less than a normal amount of it. If growth is rapid enough, blood supply after a time will be insufficient to restore used-up nutrients, especially because new vessels usually do not appear rapidly enough to keep pace with the needs of the whole expanding tumor.

The result is that a portion of an enlarging tumor may die, literally of malnutrition and oxygen lack. It is for this reason that cancers tend to ulcerate and bleed, sometimes producing thick, slimy deposits of necrotic tissue (from the Greek *nekrosis*, meaning "becoming dead") within their centers or at the periphery. Until mastectomy became a common operation less than a hundred years ago, the most dreaded complication of breast malignancy was not death but the fetid running sores it produced as a hapless woman's chest wall was digested away. This is precisely why the ancients referred to *karkinoma* as the "stinking death."

In the late eighteenth century, Giovanni Morgagni, the author of a landmark text of pathological anatomy, said of the cancer he saw in his patients and at their autopsies that it was "a very filthy disease." Even in relatively recent times, when much more was known, malignant tumors continued to be viewed as repugnant sources of self-loathing and decay, a humiliating abomination to be concealed behind euphemisms and lies. Many are the stories of women with breast cancer who withdrew from friends, secluded themselves at home, and lived their final months as recluses, sometimes even from their own families. As recently as the period of my training, just over thirty years ago, I saw a few such women who had finally been prevailed upon to come to the clinic because their situations had become intolerable. Of the several reasons we still hesitate to utter the word *cancer* in the presence of a patient or family assaulted by it, the residual heritage of its odious connections is the one most difficult for our generation to expunge.

Not enough that a rapidly growing cancer may so infiltrate a solid organ like the liver or kidney that insufficient tissue remains to perform the organ's functions effectively; not enough that it may obstruct a hollow structure like the intestinal tract and make adequate nourishment impossible; not enough that even a small mass of it can destroy a vital center without which life functions cannot go on, as some brain tumors do; not enough that it erodes small blood vessels or ulcerates sufficiently to result gradually in severe anemia, as it often does in the stomach or colon; not enough that its very bulk sometimes interferes with the drainage of bacteria-laden effluents and induces pneumonia and respiratory insufficiency, which are common causes of death in lung cancer; not enough that a malignancy has several ways by which it can starve its host into malnutrition—a cancer has still other ways to kill. Those just mentioned refer, after all, only to potentially lethal consequences of encroachment by the primary tumor itself, without its ever having left the organ where it first arose. These are the kinds of damage cancer does in its own neighborhood. But it has an additional way of killing that takes it out of the category of localized disease and permits it to attack a wide assortment of tissues far from its origin. That mechanism has been given the name *metastasis*.

Meta is a Greek preposition meaning "beyond" or "away from," and *stasis* connotes "position" or "placing." Introduced as early as the Hippocratic writings to indicate a change away from one form of fever to another, *metastasis* later came to be applied specifically to migration of bits of tumor. In modern times, this one word, *metastasis*, has come to articulate the defining feature of malignancy—cancer is a neoplasm that has the potential to go beyond its home and travel to some other place. A metastasis is, in effect, a transplant of a sample of the primary tumor to another structure or even a distant part of the body.

Cancer's ability to metastasize is both its hallmark and its most menacing characteristic. If a malignant tumor did not have the ability to travel, surgeons would be able to cure all but those that involve vital structures, which cannot be removed without compromising life. In order to travel, the tumor must erode through the wall of a blood vessel or lymph channel, and then some of its cells must become detached and pass into the flowing stream. Either individually or clumped into an embolus, the cells are then carried to some other tissue, where they implant and grow. Determined by the route of blood or lymph flow as well as other still-unclear factors, various cancers have a predilection to be deposited in certain specific organs. For example, a breast cancer is most likely to metastasize to bone marrow, lungs, liver, and, of course, the lymph nodes in the armpit, or axilla. A cancer of the prostate commonly travels to bone. Bones, in fact, along with the liver and kidney, are the most common sites for metastatic deposits, regardless of the malignancy's organ of origin.

In order to take root in a distant location, tumor cells need to be hardy enough to resist destruction while on their journey. The simple mechanical dangers of traveling through the jolting circulation complicate the possibility of being killed by the host's immune system during the course of the passage. If they survive the voyage, the cells must then establish a new home and be provided a reliable source of nutrition. This means *a priori* that the transplanted bit of cancer cannot create a viable colony on the newly reached distant shore unless it is capable of stimulating the growth of tiny new blood vessels to supply its needs.

So difficult is it to satisfy all of these requirements that very few of the migrating cells ever do manage to colonize some far-flung site. When tumor cells are experimentally injected into mice, only one-tenth of 1 percent survive beyond twenty-four hours; it is estimated that only one of each 100,000 cells entering the bloodstream lives to reach another organ, and a far smaller proportion successfully implant themselves. Were it not for obstacles such as these, massive numbers of metastases would appear as soon as a cancer becomes sufficiently large to shed many cells into the circulation.

By the twin forces of local invasion and distant metastasis, a cancer gradually interferes with the functioning of the various tissues of the body. Tubular organs are obstructed, metabolic processes are inhibited, blood vessels are eroded sufficiently to cause minor and sometimes major bleeding, vital centers are destroyed, and delicate biochemical balances are deranged. In time, a point is reached at which life can no longer be sustained.

In addition, there are less direct ways for cancer to take its toll on those in whom its growth is unchecked, and they are usually the result of the debilitation, poor nutrition, and susceptibility to infection that come with the malignant process. Nutritional depletion is so common that a term has been devised to designate its effects: *cancer cachexia.* Cachexia is derived from two Greek words meaning "bad condition," which is exactly the situation in which advanced cancer patients find themselves. It is characterized by weakness, poor appetite, alterations in metabolism, and wasting of muscle and other tissues.

Actually, cancer cachexia is sometimes present even in people whose disease is still localized and relatively small, so it is clear that factors account for it other than a tumor's gobbling up of its host's resources. Though a tumor is capable of depriving its host of some essential nutrients, the concept of parasitizing may be, in fact, a simplistic way of looking at far more complicated causes of its ability to deplete resources. Changes in taste perception, for example, and local tumor effects such as obstruction and swallowing problems sometimes contribute to inadequate intake, as do chemotherapy and X-ray treatment. Numerous studies of people with malignancies reveal various kinds of abnormalities in the utilization of carbohydrates, fats, and proteins, the causes of which are uncertain. Some tumors even seem capable of increasing a patient's expenditure of energy, thereby contributing to the inability to maintain weight. To add to the problem, certain malignancies and even some of the host's own white blood cells (monocytes) have been shown to release a substance appropriately given the name cachectin, which decreases appetite by direct action on the brain's feeding center. Cachectin is not the only such agent. It is likely that tumors of all sorts are capable of secreting various hormonelike substances which produce generalized effects on nutrition, immunity, and other vital functions that until recently were attributed to the parasitizing effects of the growth itself.

Malnutrition causes problems far beyond weight loss and exhaustion. The healthy body adapts to ordinary starvation by using fats as its main energy source, but this process is not effective in cancer, with the result that protein must be utilized. Not only does this and the lessened food intake cause muscle wasting; the decreased protein levels contribute to the dysfunction of organs and enzyme systems, and may significantly affect the immune response. There is evidence that one of the substances released by tumor cells further depresses immunity. Although this may, at least theoretically, enhance cancer growth, that untoward effect seems much less important than the fact that depressed immunocompetence, especially when magnified by chemotherapy and radiation, increases susceptibility to infection.

Pneumonia and abscesses, along with urinary and other infections, are frequently the immediate causes of death of cancer patients, and sepsis is their common terminal event. The profound weakness of severe cachexia does not permit effective coughing and respiration, increasing the chances of pneumonia and the inhalation of vomitus. The final hours are sometimes accompanied by those deep, gurgling respirations that are one of the forms of the death rattle, quite distinct from the agonal bark of a James McCarty.

Near the end, a decreased volume of circulating blood and extracellular fluid not infrequently leads to a gradual decrease in blood pressure. Even if this does not proceed to shock, it may cause organs such as the liver or kidney to fail because of chronic lack of sufficient nutrients and oxygen, although they are not directly involved with tumor. Since many people with cancer are in an older age group, the various forms of depletion often induce stroke, myocardial infarction, or heart failure. Of course, the presence of a generalized disease of metabolism, like diabetes, complicates the problems enormously.

Thus far, only those cancers have been mentioned that begin as tumors originally localized to a specific organ or tissue. A smaller group of malignant diseases have a more generalized distribution from the very beginning, or arise in multiple sites of a particular kind of tissue, specifically the blood and lymph systems. Leukemia, for example, is a cancer of the tissues responsible for the production of white blood cells, and lymphoma is a malignancy of lymph glands and similar structures. Patients with leukemia and lymphoma are particularly prone to infection, and it is a leading cause of death in those malignancies. One of the common forms of lymphoma is Hodgkin's disease.

I cannot mention Hodgkin's disease without calling attention to a remarkable accomplishment that is in many ways exemplary of the biomedical achievements of the last third of the twentieth century. Thirty years ago, virtually every patient with Hodgkin's disease died of it, unless claimed by something else in the several-year interval between diagnosis and the terminal phase. Since then, improved understanding of the way in which the disease distributes itself in the lymph glands, and its responsiveness to appropriate programs of chemotherapy and supervoltage X-ray, have resulted in five-year disease-free survival of approximately 70 percent, which is as high as 95 percent for patients whose disease is discovered when its extent is still limited; recurrence rates after this period are low and decrease with each year. Not only Hodgkin's disease but lymphomas in general are now among the most curable of all cancers.

The changed outlook for people with lymphoma is only one example of extraordinary progress in treating cancer. Another is childhood leukemia. Four out of five children with this disease have a form of it called acute lymphoblastic leukemia, previously fatal in every case; today, the five-year rate of continuous remission of acute lymphoblastic leukemia is 60 percent, and most of these youngsters will be cured. Although there have thus far been only a few other success stories of the sheer magnitude of these two, the general trend in the campaign against cancer is favorable enough to justify cautious optimism. Basic research, new ways of interpreting the clinical phenomena of disease, innovative applications of pharmacology and the physical sciences, and the willingness of informed patients to enroll in large-scale trials of promising treatments are among the reasons for the vast changes over the past few decades.

In the year I was born, 1930, only one in five people diagnosed with cancer survived five years. By the 1940s, the figure was one in four. The effect of modern biomedicine's research capacity began to make itself felt in the 1960s, when the proportion of survivors reached one in three. At the present time, 40 percent of all cancer patients are alive five years after diagnosis; making proper statistical allowances for those who die of some unrelated cause, such as heart disease or stroke, 50 percent survive at least that long. It is well known that those who reach the five-year milestone free of disease face greatly decreased odds of eventual recurrence of their malignancy. Virtually all of the progress has been made possible by a combination of earlier diagnosis and the improved treatment resulting from the factors listed in the preceding paragraph. Improved treatment and the possibility of success of the constantly appearing innovative approaches to advanced disease bring hope to today's cancer patient. Paradoxically, and sometimes tragically, that kind of hope is the very thing that has led to some of the most error-fraught dilemmas that patients and their doctors are compelled to face today.

My clinical career encompasses a period during which a realistic expectation first began to be felt in the scientific community that malignant disease would prove amenable to treatment based on an understanding of cellular biology rather than the ages-old oversimplifications of surgery. As more was learned about the cancer cell, new and increasingly effective ways were developed to combat its unchecked ravages. With the optimism born of therapeutic successes came a determined cockiness that sometimes goes beyond reason; it finds expression in the philosophy that treatment must be pursued until futility can be proven, or at least proven to the satisfaction of the physician.

The boundaries of medical futility, however, have never been clear, and it may be too much to expect that they ever will be. It is perhaps for this reason that there has arisen the conviction among doctors—more than a mere conviction, it is nowadays felt by many to be a responsibility—that should error occur in the treatment of a patient, it must always be on the side of doing more rather than less. Doing more is likely to serve the doctor's needs rather than the patient's. The very success of his esoteric therapeutics too often leads the physician to believe he can do what is beyond his doing and save those who, left to their own unhindered judgment, would choose not to be subjected to his saving.

XI

Hope and the Cancer Patient

A YOUNG DOCTOR learns no more important lesson than the admonition that he must never allow his patients to lose hope, even when they are obviously dying. Implicit in that oft-repeated counsel is the inference that a patient's source of hope is the doctor himself, and the resources he commands; thus, only a doctor has the power to offer hope, to withhold it, or even to take it away. There is a great deal of truth in such an assumption, but it is not the whole story. Beyond the medical establishment—and beyond even the capability of one's own physician, no matter his beneficence—is the power that rightfully belongs to the patient and those who love him. In this chapter and the next, I will write of people with terminal cancer, some of their kinds of hope, and how I have seen them enhanced or enfeebled—and sometimes destroyed altogether.

Hope is an abstract word. In fact, it is more than just a word; hope is an abstruse concept, meaning different things to each of us during different times and circumstances of our lives. Even politicians know its hold on the human mind, and the mind of the electorate.

Scanning my *Webster's Unabridged*, I find five separate interpretations of the meaning of the noun *hope*, and that doesn't include the synonyms. The meanings listed range from "the highest degree of well-founded expectation" to expectation that is "at least slight." In a separate entry is to be found an example of usage for *hope* as an intransitive verb, and herein may lie the crux of the matter for many patients suffering with terminal cancer: "*to hope against hope*," which the lexicographers describe as "having hope though it seems to be baseless." A physician has no greater obligation than to be sure that no hope is baseless if he has given his patient reason to believe in it.

When the *Oxford English Dictionary* is consulted, there are no fewer than sixty examples illustrating the different uses of the noun. Truly, hope springs eternal, if not necessarily in the human breast, at least in the human propensity for making a word mean "just what I choose it to mean—neither more nor less," as Lewis Carroll's Humpty Dumpty scornfully proclaimed to Alice. The meaning that hope brings is perhaps best expressed by Samuel Johnson: "Hope," wrote England's greatest authority on words, "is itself a species of happiness, and perhaps the chief happiness which this world affords."

All of the definitions of hope have one thing in common: They deal with the expectation of a good that is yet to be, a perception of a future condition in which a desired goal will be achieved. In a very perceptive passage in his book *The Nature of Suffering*, the medical humanist Eric Cassell writes with great sensitivity of the meaning of hope in times of serious illness: "Intense unhappiness results from a loss of that future—the future of the individual person, of children, and of other loved ones. It is in this dimension of existence that hope dwells. Hope is one of the necessary traits of a successful life."

I would argue that of the many kinds of hope a doctor can help his patient find at the very end of life, the one that encompasses all the rest is the belief that one final success may yet be achieved whose promise vanquishes the immediacy of suffering and sorrow. Too often, physicians misunderstand the ingredients of hope, thinking it refers only to cure or remission. They feel it necessary to transmit to a cancer-ridden patient, by inference if not by actual statement, the erroneous message that it is still possible to attain months or years of symptom-free life. When an otherwise totally honest and beneficent physician is asked why he does this, his answer is likely to be some variation of, "Because I didn't want to take away his only hope." This is done with the best of intentions, but the hell whose access road is paved with those good intentions becomes too often the hell of suffering through which a misled person must pass before he succumbs to inevitable death.

Sometimes it is really to maintain his own hope that the doctor deludes himself into a course of action whose odds of success seem too small to justify embarking on it. Rather than seeking ways to help his patient face the reality that life must soon come to an end, he indulges a very sick person and himself in a form of medical "doing something" to deny the hovering presence of death. This is one of the ways in which his profession manifests the entire society's current refusal to admit the existence of death's power, and perhaps even death itself. In such situations, the doctor resorts to a usually ineffective delaying action that utilizes what has been called by a leading physician of the generation just past, William Bean of the University of Iowa, "the busy paraphernalia of scientific medicine, keeping a vague shadow of life flickering when all hope is gone. This may lead to the most extravagant and ridiculous maneuvers aimed at keeping extant certain representative traces of life, while final and complete death is temporarily frustrated or thwarted."

Dr. Bean was referring here not just to the respirators and other end-of-life artificialities but to the whole gamut of stratagems whereby we attempt to turn our eyes away from the fact that nature always wins. This is the baseless hope that contradicts expectation; it was the kind of "hope against hope" to which I succumbed a few years ago when my own brother was diagnosed with widely metastatic intestinal cancer.

Harvey Nuland was a healthy sixty-two-year-old man who occasionally visited a doctor when he was concerned about some specific symptom, but otherwise was not inclined to undergo medical surveillance. He carried ten to fifteen extra pounds on his compact frame, but he was hardly obese. His work as an executive partner in a large New York accounting firm was a source of enormous gratification for him, although it demanded long hours and great responsibility—perhaps *because* it demanded long hours and great responsibility. The focus of my brother's life was not his work, however. Harvey's happiness was invested in his family. He had not married until his late thirties and did not become a father until he was past forty. That, and the disjointed nature of our lives as he and I were growing up, may have been the reasons that the closeness of his family became the paramount fact of his life, almost as though it had been sanctified by having come as such a late blessing.

One morning in November of 1989, Harvey phoned to tell me that he had been having bowel irregularities and pain for a few weeks, and the previous afternoon had been found by his doctor to have a mass on the right side of his abdomen. There were to be definitive X rays later in the day, and he wanted me to be aware of what was going on. He tried to speak matter-of-factly, but we had been through far too much together for me to be fooled. Neither was he taken in by some reassuring words I managed to come up with. Even this most guileless of men was not to be sweet-talked out of his anxiety. We saw through each other, as brothers usually do, but only I knew just how bad his diagnosis was likely to be. A painful mass in a sixty-two-year-old man with bowel problems and a family history of intestinal cancer will almost certainly prove to be due to a partially obstructing malignant tumor—and one that is probably too far advanced for effective treatment.

The X rays confirmed my fears, and Harvey was admitted to a large university medical center. He chose it because his work had brought him into contact with a senior member of its gastroenterology staff. The surgeon I had recommended was away at a national meeting, and it was felt that impending completeness of the obstruction demanded urgent intervention. Accordingly, the operation was done by a man not personally known to me but highly praised by the gastroenterologist. Harvey was found to have a very large intestinal cancer that had invaded the tissues around his right colon and virtually all the draining lymph nodes. The tumor had deposited clumps of itself on numerous surfaces and tissues within the abdominal cavity, metastasized to at least half a dozen sites in the liver, and bathed the whole murderous outburst in a bellyful of fluid loaded with malignant cells—the findings could not have been worse. All of this had followed on a mere few weeks of symptoms.

Somehow, the surgical team managed to remove the part of the bowel in which the tumor had originated, so that Harvey's obstruction was circumvented. Masses of cancer had to be left behind—in numerous tissues and in the liver. As Harvey recovered from the operation's assault, I grappled with the twin issues of truthfulness and treatment. The decisions were mine to make, because it was clear that my brother would do as I recommended.

But how was I to be objective in trying to make clinical judgments for my own blood? And yet, I could not avoid my responsibility by pleading the emotionality of a kid brother who knew that his first childhood friend was going to die. To do that would have constituted a kind of abandonment not only of Harvey but also of his wife, Loretta, and their two college-age children.

There was no likelihood of guidance, or even understanding, from Harvey's doctors, who had by then shown themselves to be untouchably aloof and self-absorbed. They seemed too distanced from the truth of their own emotions to have any sense of ours. As I watched them strutting importantly from room to room on their cursory rounds, I would find myself feeling almost grateful for the tragedies in my life that had helped me to be unlike them. Decades of observing the highly trained university specialists who are my colleagues had persuaded me of the sensitivity of most and the isolation of the relatively few. In this place, the few seemed to be in charge of setting the scene.

With this burden on my shoulders, I made a series of mistakes. That I made them with what seemed like the best of intentions does not mitigate how I feel about them in retrospect. I became convinced that telling my brother the absolute truth would "take away his only hope." I did exactly what I have warned others against.

Harvey had very blue eyes. So do I and so do all four of my children. Our blue eyes are an inheritance from my mother. Every time I visited my brother during the first of those three long postoperative weeks in the hospital, his pupils were constricted to pinpoints by morphine or some other narcotic, necessitated by the unremitting pain of his ribs-to-pubis incision. Although very nearsighted, he rarely wore his glasses during that time, and I saw in those wondrously blue eyes a look that had not been there since we were two kids playing stickball in the Bronx in the few hours free from our after-school jobs. Sickness had somehow restored to Harvey his innocence of early adolescence, and his trust. He seemed a boy again, this big brother to whom I had so often in my life turned for counsel and help. And I, in my vibrant health, remained a grown man. I resolved during those postoperative days that I would protect my brother from the anguish suffered by those who know there is no hope for cure. In retrospect, I now realize that I was trying to protect myself as well.

I knew of no form of chemo- or immunotherapy that might deter so advanced a cancer from its course. In New Haven, I "discussed the case" (a euphemism for what I really did, which was to scrape the brains of oncologists in my search for a miracle) with colleagues. Several times, I tried to talk things over with Harvey's doctors, which I found an exercise in frustration and a lesson in medical arrogance. I heard about an experimental new treatment using an unusual combination of two agents in a way never tried before. One of the drugs, 5-fluorouracil, interferes with the metabolic processes of cancer cells, and the other, interferon, exerts antitumor effects in ways not yet completely understood. The 5-fluorouracil–interferon program had decreased tumor bulk in eleven of nineteen patients in the only group of any size that had yet been tested, but it had cured no one. The small number of treated patients had suffered an assortment of major toxic side effects, and there was even one chemotherapy-induced death.

I sought out the doctor at Harvey's hospital who had experience with the drug combination. I let my instincts as a brother overwhelm my judgment as a surgeon who has spent his career treating people with lethal disease. What could have made me believe that a unique medical coincidence had somehow occurred to solve what my rational mind knew was insoluble? Could I really have thought that a potential cure or even a reasonable palliative had somehow magically appeared just at the moment when my brother was found to have a cancer I knew to be beyond any treatment? Looking back on it, I'm not sure what I thought—I seem only to have been motivated by my inability to tell Harvey the truth of his prognosis.

I could not face my brother and speak the words that should have been said; I couldn't tolerate the immediate burden of hurting him, and so I exchanged the possibility of the comfort that may come with an unhampered death for the misconceived "hope" I thought I was giving him.

I had looked into those boyishly trusting blue eyes and seen my brother asking me for deliverance. I knew I was not able to give it, but I knew also that I could not bring myself to deprive him of the hope that I would somehow find a way. I told him about the cancer in his colon and the metastases in his liver but chose not to reveal the extent of the deposits elsewhere or the significance of the fluid. At no time did I ever consider sharing with him what I knew to be the virtually certain prognosis that he would not survive till summer. In every way, I had returned to the misconceived paternalistic dictum of the professors who taught me a generation ago: "Share your optimisms and keep your pessimisms to yourself."

In all of this, I took my cues from Harvey's eyes and his words. No one who has treated cancer patients will ever discount the power of the subconscious mechanism we call denial, which is both friend and enemy of a person seriously ill. Denial protects while it hinders, and softens for a moment what it eventually makes more difficult. As much as I applaud Elisabeth Kübler-Ross's attempt to categorize a sequence of responses to the diagnosis of mortal illness, every experienced clinician knows that some patients never, at least overtly, progress beyond denial; many others retain large elements of it right to the end, in spite of every effort that might be made by a physician to clarify each issue as it arises. Explanations of the forcefulness of denial's influence are themselves often denied. Harvey Nuland had a first-class mind and two perfectly good ears, not to mention the keen degree of insight common in those accustomed to adversity, and yet—again and again—I was to be taken aback by the magnitude of his denial, until near his last days. There was something in him that refused the evidence of his senses. The clamor of his wish to live drowned out the pleadings of his wish to know.

Denial is one of two factors that immeasurably complicate our best intentions when, as physicians or the beloved of a dying person, we seek to enlist him as a full participant in choices that must be made in the days remaining. Few dying people with a clear understanding of the inevitability of their disease process are willing to suffer through heroic and debilitating attempts to fight off an end that seems close. It is in the "clear understanding of their disease process," however, that reason and logic sometimes founder, and denial is a major element that stands in the way. Denial is a significant factor, for example, in the surprising frequency with which dying people refuse to confront the nearness of circumstances they anticipated when, while still healthy, they signed advance directives prohibiting major resuscitative efforts. When the chips are down, almost no one wants his life to end, and one good way for the conscious mind to avoid it is for the unconscious mind to deny that it is about to happen.

The other hindrance to full participation is the refusal of many patients to exercise their right to independent thought and self-determination—in other words, their control. The psychoanalyst and legal scholar Jay Katz has used the term *psychological autonomy* to denote this right of independence. Many a patient worn down by the ravages of illness or overwhelmed by the immediacies of a dire situation will be unwilling or emotionally unable to use his autonomy. The need to be cared for and to be relieved of responsibilities is not easily dealt with under such circumstances, and it may lead to wrong decisions. But the problem may be lessened if both patient and caregivers reflect on it together. When this is done, a dying man will sometimes decide that he wants to participate much more actively than he thought he could. If he does not, this, too, must be respected.

In trying to do the right thing for Harvey, I became what he wanted me to be, and in so doing fulfilled both his fantasy of me and my own: the smart kid brother who had gone off to medical school and grown up to be the all-knowing and quite omnipotent medical seer. I could not deny him a form of hope that he seemed to need. I would marshal the forces of cutting-edge medicine and rescue him from the brink of death. This is every doctor's most pervasive near-conscious self-image, and my brother's eyes persuaded me to succumb to it. Had I been wiser, or consulted disinterested colleagues who knew me well, I might have understood that my way of giving Harvey the hope he asked for was not only a deception but, given what we knew about the toxicity of the experimental drugs, an almost certain source of added anguish for all of us.

Harvey required three further hospitalizations in the ten months of life that were left to him after his operation. He was admitted to monitor the initiation of the chemotherapy, and near the end he had to return when growing tumor deposits obstructed his intestine again, this time completely. The obstruction spontaneously opened just enough to let him take sufficient liquids by mouth to avoid reoperation, but not to maintain even the dwindling state of his previous nutrition. As difficult as was this last period in the hospital, it was the one before it that has left the most tormenting memories.

Harvey's son, Seth, had taken a year's leave of absence from school to work on a kibbutz in Israel, but came home to be his father's primary caregiver because Harvey insisted that his wife, Loretta, not give up her full-time job at a local community college. Seth phoned me one Friday evening to tell me that Harvey had been lying on a stretcher outside the hospital's emergency room for two days, suffering the effects of severe drug toxicity and passing in and out of coma. He, his sister, Sara, and Loretta had been taking turns at his side, though he often did not know they were there. No bed was available on any floor in the entire building. The toxic effects of the drugs—nausea, diarrhea, depression of the bone marrow's ability to make white blood cells—were a problem from the start but had lately become increasingly unmanageable. Obviously, things were now out of control. The professor who was Harvey's oncologist had gone away for the weekend and his training fellows seemed uninterested or unable to do much beyond ordering an intravenous drip.

When I arrived at the hospital the next morning, I found every cubicle in the chaotic emergency room occupied. Crowded into the narrow corridor outside were at least seven stretchers, on which lay some of the sickest people I had ever seen packed into one small area, and almost all of them appeared to have AIDS or advanced malignancy. As I picked my way carefully through the narrow spaces between patients and their anxious families and friends, I looked up and saw my nephew standing disconsolately next to a gurney on which lay his unconscious father. At the foot of the gurney sat my niece, hunched over and staring at the floor. She looked in my direction and tried to give me a wan smile, but tears began streaming down her face.

During all of those three days when Harvey was passing in and out of stupor in that cluttered hospital corridor, his temperature had been ranging between 102 and 104 degrees. In spite of valiant efforts by the overwhelmed nurses attempting to provide at least a modicum of care for everyone, and the help given Harvey by his wife and children, he had lain for long periods of time in the liquid diarrhea that periodically poured spontaneously out of him in response to the ravaging effect of the drugs on his intestinal tract. Even his periods of consciousness were not completely lucid, and most of the time he was uncertain of either his whereabouts or his condition.

I spoke to the harried resident physician who had been calling the admitting office over and over to try to place some of her sickest patients, and she agreed to make one more attempt, happy at the opportunity to use my medical connection to help at least one of them get into a real bed. An impressionable clerk must have been on duty, because the strategy worked—within two hours, Harvey was upstairs on one of the nursing floors. As we wheeled him toward the elevator, I sneaked a last guilty look in the direction of the space alongside the one we were vacating, where an exhausted boy not much older than my nephew was hovering over a blanket-covered stretcher. He was speaking softly to his shivering friend, another young man close to death from AIDS.

Harvey paid a high price for the unfulfilled promise of hope. I had offered him the opportunity to try the impossible, though I knew the trying would be bought at the expense of major suffering. Where my own brother was concerned, I had forgotten, or at least forsaken, the lessons learned from decades of experience. Thirty years earlier, when there was no chemotherapy, Harvey would probably have died at about the same time that he eventually did, of the same cachexia, insufficiency of the liver, and chronic chemical imbalance, but his death would have been without the added devastation of futile treatment and the misguided concept of "hope" that I had been reluctant to deny him and his family, as well as myself. When I have explained the high frequency of dangerous toxicity of certain desperate forms of treatment whose likelihood of success is remote, some of my advanced cancer patients have wisely chosen to do nothing, and found their hope in other ways.

By the time Harvey recovered from this nearly lethal episode, his liver metastases, which had initially responded to the new treatment by a shrinkage of 50 percent, were once more enlarging. Because of this and the fact that the other areas of tumor had never stopped growing, it was clear that there was no longer any justification for the continuing chemotherapy. He returned home to die.

It was at this point that the local hospice was called in. I had been a board member of the Connecticut Hospice, and many of my terminally ill cancer patients had benefited from the care that these devoted nurses and doctors provide. Their goal is comfort, and their concept of comfort includes the totality of the life of patients and their families. The local hospice set to work immediately, showing Loretta how she could organize the household in ways that would minimize Harvey's distress. Seth was taught to administer medications for pain and nausea, and learned useful techniques to help his father get around the house.

One additional hospitalization became necessary when continued growth of the cancer finally obstructed the intestine. So many areas of small bowel were tethered into the encroaching tumor mass that no surgery was possible. Just when the situation seemed to have reached its conclusion, the gut spontaneously opened just enough so that Harvey could return home. This time, I asked my original choice of surgeon to take over, and I will ever be grateful to him for restoring to all of us a sense of commitment and kindness, as well as common sense.

Even with the frequent hospice visits and the selfless care given by Seth, who had by then become Harvey's constant companion and his nurse, the pain and increasing weakness were difficult to manage. The narrowness of the intestinal passage prevented retention of any but a little nourishment; medication had to be given by suppository. Harvey had already lost a great deal of weight, but now his cachexia rapidly worsened.

When I visited, Harvey and I would sit together on the couch and try to keep each other's spirits up. A few times, when we were briefly alone, we talked about Loretta and the kids and how things would be after he was gone. Sometimes we spoke not of the future lost to him but of the long-ago past that seemed like yesterday, when we were boys in the Bronx speaking Yiddish to Bubbeh. Gone were the petty irritations and occasional conflicts that arise when two strong-willed brothers marry and their lives go off in different directions. It comforted me, in those last weeks, to remind Harvey of the several troubled times I had experienced decades before, when he was the one person who knew how to help me—more than twenty years earlier, I had left all that mattered in my life and traveled to a distant cheerless shore, from which I returned only because he never doubted that I would. No matter the remoteness that had sometimes come between us, neither had ever doubted the other's love, but now it became important for each of us to say it. I kissed him each time I left to return to New Haven. The last time was two days before his long travail ended quietly in the bed he and Loretta had shared for so many years.

During the several days after the funeral, I went each morning with Seth and Sara to recite the mourner's prayer, the Kaddish, at the synagogue where less than two years earlier I had gone to a testimonial dinner honoring Harvey at the conclusion of his term as president of the congregation. I knew the words of the prayer by heart because I had needed them often since that cold December morning half a century ago when Harvey and I stood together at our mother's open grave, saying them for the first time.

In this high-tech biomedical era, when the tantalizing possibility of miraculous new cures is daily dangled before our eyes, the temptation to see therapeutic hope is great, even in those situations when common sense would demand otherwise. To hold out this kind of hope is too frequently a deception, which in the long run proves far more often to be a disservice than the promised victory it seems at first.

Mine is not the first voice to suggest that as patients, as families, and even as doctors, we need to find hope in other ways, more realistic ways, than in the pursuit of elusive and danger-filled cures. In the care of advanced disease, whether cancer or some other determined killer, hope should be redefined. Some of my sickest patients have taught me of the varieties of hope that can come when death is certain. I wish I could report that there were many such people, but there have, in fact, been few. Almost everyone seems to want to take a chance with the slim statistics that oncologists give to patients with advanced disease. Usually, they suffer for it, they lay waste their last months for it, and they die anyway, having magnified the burdens they and those who love them must carry to the final moments. Though everyone may yearn for a tranquil death, the basic instinct to stay alive is a far more powerful force.

About ten years ago, I cared for a man whose despair and paralyzing fear of treatment drove him to seek hope in other than medical efforts. He gave up the possibility of cure and became reconciled to his death, or at least determined that if miracles were to occur, they would come from within himself and not from some enthusiastic oncologist.

Robert DeMatteis was a forty-nine-year-old attorney and a political leader in a small Connecticut city, and he was terrified of doctors. Fourteen years earlier, I had treated him for extensive injuries following an automobile accident, and I was astonished at his inability, during the period of hospitalization, to tolerate so much as the most minimal discomfort or even the possibility that it might occur. The fact that his wife, Carolyn, was a nurse diminished not one iota of the apprehension that visibly mounted in him at the mere approach of a white-coated figure. Carolyn once told me that he used to insist she change out of her uniform while still at the hospital where she worked, because the sight of it in his own home made him anxious.

Bob was the sort of man to whom no one gave orders. He seemed to take pride in being obstinate, and one of the manifestations of that trait was an arrant disregard for his health. He had ignored not only his health but everything else about his body except its enormous appetite for good food. At a height of five feet eight inches, Bob DeMatteis weighed 320 pounds. To his family, a large circle of friends, and to those many townspeople who came to him for help in solving a problem, this misanthropic-looking fellow was a warmhearted, gregarious man. Nevertheless, the first sight of his Harry the Horse build and the scowl above it had the effect of intimidating the faint of heart. He was as intense in his loyalties as he was in his conflicts, a man accustomed to deference. The menacing quality of his low-pitched, gravelly voice made even tenderness sound like a growl.

Bob seemed hardly the sort of man who would cringe at the sight of a young woman carrying a hypodermic syringe. He joked about his fear, but it sometimes stood in the way of proper care, and more than once during the trauma hospitalization it prevented me from treating his wounds in an optimal manner.

With all of these fourteen-year-old recollections as background, I was not pleased when Bob's internist called me one mid-May afternoon. Bob had been admitted that morning after passing a large quantity of fresh blood rectally, and was being transfused on the medical floor. When I saw him he provided an interesting clue that made me think he had actually been oozing very small amounts of blood for the previous several months, before the present sudden hemorrhage. He said that he had been experiencing a gradually worsening abdominal discomfort since February. He also described a subtle but definite alteration in the odor of his stools. The color had not changed but the new smell was unmistakable—it was produced by the presence of blood. A month earlier, when Carolyn had finally dragged him to his internist, protesting all the way, a series of X rays were done, which showed a superficial erosion of the duodenum but no ulcer. Some thickening had been noted at the ileocecal valve, which is the point where the small intestine enters the colon. Bob was reassured that no tumor was seen.

The rapid bleeding stopped within a few hours of Bob's admission to the Yale–New Haven Hospital, and it was possible to carry out a complete evaluation of the gastrointestinal tract. Attention was focused on the colon rather than the upper portion because of the peculiar thickening seen on X rays as well as a few of the physical findings. We were not surprised when the fiberoptic visualizing instrument called a colonoscope revealed not a thickening but a tumor at the ileocecal valve.

Predictably, Bob reacted with hysteria to the news that he would need an operation, to which he flatly refused to agree. When he calmed down a little, he began to growl and complain and even swore a bit, but the patient urging of his wife finally brought him to consent. I don't think I have ever taken a more frightened person to the OR. I always try to be at a patient's side while the anesthesia is induced, so that I can speak to him and hold his hand, but being with Bob was a totally new kind of experience. Afterward, I had to massage my fingers for a few minutes before beginning to scrub, because he seemed to have squeezed the blood out of them by the time he reluctantly allowed himself to go under.

The operative findings were a shock. Expecting to see a relatively small tumor that had ulcerated just enough to bleed, what I encountered was nothing less than (and here I quote from the pathology report) a "poorly differentiated primary adenosquamous carcinoma arising in cecum adjacent to ileocecal valve, exhibiting transmural [through the wall] invasion into peri-colic fat, extensive lymphatic and vascular involvement and metastases to 8 of 17 lymph nodes." The center of the tumor was necrotic and deeply ulcerated, which accounted for the episode of brisk bleeding.

Although there was as yet no visible evidence of distant metastasis, Bob's cancer was obviously very aggressive. With such extensive invasion of the blood vessels and lymph channels, the presence of large numbers of cancer cells in the general circulation was a certainty. It was almost equally certain that there were already some deposits in the liver that were still either microscopic or simply too deep to feel. It would be only a matter of time before they gave some overt evidence of their presence. Bob's prognosis was terrible.

Bob DeMatteis was as blunt and direct as he looked, and he had a fine ear for evasion. He demanded to know exactly what he was facing, page and number—no details were to be left out. My behavior with Harvey notwithstanding, I have always tried to set the stage for patients to ask for full disclosure, and I welcomed his questions even though I anticipated that I might regret the unadorned candor he seemed to demand. I took him at his word, expecting him to break down into hysteria and lapse later into profound depression. It never happened.

There was no emotional outburst—not a bit of it. Calm, reason, and acceptance took its place. As early as their period of courting, Bob had told Carolyn (and to this day she does not know why) that he did not expect to see his fiftieth birthday, and his prophecy was about to be fulfilled. At the end of our first postoperative conversation, Bob knew he was going to die of his cancer, and he planned to let it happen without interference. He was not a religious man, but he had an abiding faith in himself, which at this point became the gyroscope that stabilized his remaining time.

Bob reckoned without the oncologists. In view of (by my lights, *in spite of*) the advanced state of disease, the option of consultation was given to him after his wife and internist had initiated the idea. Neither he nor I had any enthusiasm for it, but he agreed to speak to an oncologist, if for no other reason than to placate Carolyn, who was determined that no possibility go unexplored. At that point (and even now, more than a decade later), I had never had a single experience in which an oncology consultation did not result in a recommendation to treat, unless the disease was so early that surgery had definitely cured it. Bob's case was no exception, and Carolyn prevailed on him to accept the course of therapy being offered.

The chemotherapy had to be delayed for a reason almost unique to very obese people: The enormous layer of fat under Bob's skin was much too thick for me to consider closing it at the time of the operation, lest a hidden abscess develop in its depths. In order to guarantee clean healing, I was forced to leave it open to seal from bottom to top, which held up the drug treatments for an extended period. By the time they could be started, the liver metastases of this rapidly expanding tumor had grown large enough that they could be identified with radioactive isotope studies.

Before embarking on a course of therapy, the oncologist met with Bob to have what he (the oncologist) later described to me in a letter as "an extended and frank discussion," during which he "detailed the extent of metastatic disease and told him that if chemotherapy was not effective he might rapidly go downhill and expire within the next three to six months." He reported that Bob "was very appreciative of the frank discussion, and has a cautiously optimistic but realistic attitude."

By this time, Bob had regained the twenty pounds lost since his operation, and he was without any symptoms. In fact, he was feeling remarkably well. He understood that the drugs could not cure but were to be used "in an adjuvant or preventive fashion," as the oncologist put it. I doubt that Bob expected even that; more likely, he was just going through the motions for the sake of Carolyn and their twenty-year-old daughter, Lisa. The treatment was begun.

Within two weeks, Bob had developed high fevers and constipation alternating with diarrhea. The skin between his corpulent buttocks was tender and reddened by the corrosive effect of the loose stools. The chemotherapy had to be stopped. About this time, narcotics were becoming necessary to control the onset of pain being caused by the enlarging liver metastases. Soon, Bob was no longer able to go to his office.

With shocking speed, the metastases grew bulkier and Bob became jaundiced as his liver was progressively replaced by cancer. Evidence appeared of a mass of tumor in his pelvis, and his legs were soon swollen with the edema that results when veins draining the lower body are blocked by cancer pressing on them. After a time, Bob could barely make his way around the house. Because Carolyn was working, Lisa stayed at home to care for him. As she would tell me years later, "We spent many long nights talking about ourselves and each other. As close as we had been before then, we became even closer as those last few months went by."

Early on Christmas Eve, I made a house call. The DeMatteises lived in a wooded area in the hills above the outskirts of the city of whose political life Bob had so long been a driving force. Snow had begun to fall a few hours earlier, as though to honor the Christmas wish of a dying man. To Bob, this holiday had always been symbolized in an image of early nineteenth-century Dickensian joviality in which he placed himself at the center of a convivial cameraderie of festive joyousness. On this night every year since their marriage, the house had been filled with guests of every conviction and stripe, whose only shared criterion for invitation was that their host enjoyed being with them. He was at his best in a crowd, and the more high-spirited, the better. In such a company, his heart swelled and he became as large in spirit as he was in form. Even his habitual scowl absented itself from the merriment. At Christmas, Bob DeMatteis was Mr. Fezziwig and the transformed Scrooge all in one. It was his custom, in fact, to recite—not read, but recite from memory—*A Christmas Carol* to Lisa and Carolyn each year just as the holiday was about to begin. It did not surprise me to learn that Dickens was his favorite author, and this story was his favorite work of Dickens.

Bob was determined that his last Christmas would be no different from those that had come before. When the courageously smiling Carolyn opened the door, I stepped across the threshold into a house prepared for the happiest of parties. The table was set for some twenty-five people, the decorations were up, and the base of the beautifully lighted tree was hidden by piles of gifts. The guests would not begin to arrive for at least an hour, so Bob and I had plenty of time to discuss the reason for my visit. I had come to talk about hospice—I wanted its benefits for Bob. There were limits to what Lisa could do unaided now that her father's condition was worsening daily.

We sat alongside each other on the side of Bob's rented hospital bed, and after a while I took one of his hands in mine. Doing that made it somehow easier for me to speak. We were two men of the same age with quite different experiences of life, and one of us had almost used up his future. But in the short time left to him, Bob was able to see a form of hope that was his alone. It was the hope that he would be Bob DeMatteis to his last breath, and that he would be remembered for the way he had lived. Keeping this last Christmas in the best possible way was an essential part of fulfilling that hope. Then, he told me, he would be ready for the hospice nurses to take him into his final days.

As I said good-bye to this uncommon man who had found a courage I had not dreamed possible for him, I was the one whose throat choked closed. Bob was becoming impatient to start the laborious process of dressing before his guests arrived, and I was a reminder of what lay before him when the party ended. As I prepared to step out into the snowy night, he called after me from the bedroom, cautioning me to be careful on the slippery hills: "It's dangerous out there, Doc—Christmas is no time to die."

Bob made it all work that evening. He had Carolyn turn the rheostats down so that his guests would not see the full depth of his jaundice in the dimmed light. At dinner, he sat at the head of the happy, noisy table and pretended to eat, although he was long past being able to take sufficient food to get proper nourishment. Every two hours during the course of the long evening, he agonizingly dragged himself into the kitchen so that Carolyn might give him a shot of morphine to control his pain.

When all the guests had said their good-byes—so many friends of long years and decades never to be seen again—and Bob was back in bed, Carolyn asked him how the evening had been. To this day, she remembers his exact words: "Perhaps one of the best Christmases I ever had." And then he added, "You know, Carolyn, you have to live before you die."

Four days after Christmas, Bob was enrolled in the hospice home-care program, and not a day too soon. In addition to nausea and vomiting and the pain of his liver and pelvic tumor masses, he was now having high fevers. On New Year's Eve, he had a temperature of 106. His watery diarrhea was at times beyond control and frequently caught him unawares. Though it seemed impossible for the situation to worsen, it did. Finally, on January 21, Bob agreed to be admitted to the inpatient building of Connecticut Hospice in Branford. By then his liver, which in a normal state should not have extended lower than his rib margin, could be felt (even through the still-thick abdominal wall) ten inches lower than that. It was hugely enlarged, and almost all of it was cancer. In spite of the advanced degree of malnutrition, the hospice admission note records that "He was still massively obese."

Although reluctant to give in, Bob admitted to being immensely relieved on entering the inpatient facility. His underlying anxiety and restlessness had again become a problem, requiring heavy doses of tranquilizers in addition to the morphine. He was able to take only limited amounts of liquid by mouth; after his admission, he seemed to weaken by the hour. Still he persisted in struggling to get up to urinate, and he made ineffectual attempts to walk. No matter his acceptance of death, he seemed unable to let go of life.

On the afternoon of Bob's second day at the hospice, he suddenly became even more agitated than before. Carolyn and Lisa began to cry because they did not have it in their power to help him when he said that he wanted to die at that moment—immediately. As he stared pleadingly at them, he opened out his still-rotund arms and drew the two women close to him in the old protective embrace they knew so well from many times past. Holding his family together that way, he begged of them. "You have to tell me it's okay to die. I won't until you tell me it's okay." He would accept nothing less than their permission, and only when they gave it did he become calm. A few moments later, he turned to Carolyn and said, "I want to die." And then, his voice a whisper, he added, "But I want to live." After that, he became quiet.

Bob was stuporous most of the next day. By afternoon, he had not spoken, but Carolyn believed he could still hear her voice. She was speaking softly, telling him how much his life had meant to them, when all at once his face broke out in a huge smile, as if through closed eyes he was seeing some glorious thing. "Whatever it was he saw," Carolyn later told me, "it must have been beautiful." Five minutes later, he was dead.

The funeral was huge, almost a public event in Bob's city. The mayor was there and an honor guard of police met his coffin at the church. He was buried with a letter of good-bye in the pocket of his suit, from Lisa. As the cherry-wood casket was being lowered into the grave, Carolyn's uncle noticed that its lid was smudged by a small stain, where Lisa's tears had fallen on it.

Bob is buried in a Catholic cemetery about ten miles from my home. There are no monuments in those rolling hills of well-tended grave sites, as though to affirm each person's equality in death; only footstones identify the resting places. I went to visit Bob's grave during the time I was writing these last few pages, to pay homage to a man who had found a new meaning in his life when he knew he was soon to die. He had taught me that hope can still exist even when rescue is impossible. I would somehow forget his lesson when my brother fell ill a decade later, but that does not diminish its truth.

Carolyn had told me that while he was still able, Bob had arranged to have his favorite words from his favorite work of Dickens inscribed on his grave marker, but still I was unprepared for their effect when actually seen. Engraved across the granite face of the footstone was the epitaph by which Bob DeMatteis had chosen to be remembered: "And it was always said of him that he knew how to keep Christmas well."

XII

The Lessons Learned

RABBIS OFTEN END a memorial service with the sentence, "May his memory be for a blessing." It is a specific formula of words that is not familiar to the non-Jews who are present when it is said, and I have listened in vain for it in churches. Though it expresses what is obviously a universal wish, this simple thought deserves more frequent pondering by all of us, and not only in houses of worship.

The hope that brought a measure of peace to Bob DeMatteis was to be found in the memory he could create and in the meaning his life would have for those left after he was gone. Bob was a man who lived with the constant awareness that one's existence is not only finite but always in danger of ending unexpectedly. Therein lay the seed of that awful anxiety induced by things medical, but therein also lay the germinal focus of his acceptance when the final illness announced itself.

The greatest dignity to be found in death is the dignity of the life that preceded it. This is a form of hope we can all achieve, and it is the most abiding of all. Hope resides in the meaning of what our lives have been.

Other sources of hope are more immediate, but some of them will be impossible to attain. In my medical practice, I have always assured my dying patients that I would do everything possible to give them an easy death, but I have too often seen even that hope dashed in spite of everything I try. At a hospice too, where the only goal is tranquil comfort, there are failures. Like so many of my colleagues, I have more than once broken the law to ease a patient's going, because my promise, spoken or implied, could not be kept unless I did so.

A promise we can keep and a hope we can give is the certainty that no man or woman will be left to die alone. Of the many ways to die alone, the most comfortless and solitary must surely take place when the knowledge of death's certainty is withheld. Here again, it is the "I couldn't take away his hope" attitude that is so often precisely how a particularly reassuring form of hope is never allowed to materialize. Unless we are aware that we are dying and so far as possible know the conditions of our death, we cannot share any sort of final consummation with those who love us. Without this consummation, no matter their presence at the hour of passing, we will remain unattended and isolated. For it is the promise of spiritual companionship near the end that gives us hope, much more than does the mere offsetting of the fear of being physically without anyone.

The dying themselves bear a responsibility not to be entrapped by a misguided attempt to spare those whose lives are intertwined with theirs. I have seen this form of aloneness, and even unwisely conspired in it, before I learned better.

As my grandmother became no longer able, Aunt Rose gradually took over the management of our household and the mothering of its two boys. Even the matriarchal role in the extended family fell to her as Bubbeh relinquished it year by year. Early each morning, Rose went off to stitch dresses for a garment manufacturer on Thirty-seventh Street, and ten hours later she arrived home to clean the house and prepare dinner. Old World Jews did not eat lightly, and our evening meal was the result of a great deal of hard work. I am a long distance and a long time from 2314 Morris Avenue, but yesterday's memories of Thursday evenings are very clear, when Aunt Rose scrubbed and cleaned every corner of the apartment in preparation for the Sabbath, finally falling into bed near midnight, drained of energy. At six next morning, she was up again and off to work.

Rose did what she could to be brusque, but her manner was transparent. She had a pair of those blue eyes that signified our small band, and the twinkle in them was as sure to follow after an outburst of anger as is the inevitable sunny aftermath of a brief summertime shower. She was a sucker for a hug, and as we grew older, her need to appear stringently unrelenting in her expectations of her two boys slowly let itself be recognized as the love it really was. Though Harvey and I could usually tease her out of the censure that she never hesitated to express at the less admirable aspects of our behavior, we nevertheless feared her disapproval, which in my case usually took the form of denunciations, often in a colorful Yiddish, of my entire worldview and character. Aunt Rose was my little shtetl-bred superego. Harvey and I adored her.

During my second year of surgical residency, when Rose was in her early seventies, she experienced a gradual onset of generalized itching all over her body, and after a while an enlarged lymph gland appeared in her armpit. Biopsy revealed an aggressive lymphoma. She was treated by a kind and understanding hematologist who achieved an excellent remission using one of the early chemotherapy agents, chlorambucil. When after a few months the disease recurred and Rose began to weaken, Harvey and I, with the agreement of our cousin Arline, colluded to convince the hematologist that she must not be told her diagnosis.

Without perhaps even realizing it, we had committed one of the worst of the errors that can be made during terminal illness—all of us, Rose included, had decided incorrectly and in opposition to every principle of our lives together that it was more important to protect one another from the open admission of a painful truth than it was to achieve a final sharing that might have snatched an enduring comfort and even some dignity from the anguishing fact of death. We denied ourselves what should have been ours.

Although there was no doubt that Rose knew she was dying of cancer, we never spoke of it to her, nor did she bring it up. She worried about us and we worried about her, each side certain it would be too much for the other to bear. We knew the outlook and so did she; we convinced ourselves she didn't know, though we sensed that she did, as she must have convinced herself we didn't know, though she must have known we did. So it was like the old scenario that so often throws a shadow over the last days of people with cancer: we knew—she knew—we knew she knew—she knew we knew—and none of us would talk about it when we were all together. We kept up the charade to the end. Aunt Rose was deprived and so were we of the coming together that should have been, when we might finally tell her what her life had given us. In this sense, my Aunt Rose died alone.

This terrible solitude is the subject of Tolstoy's story 'The Death of Ivan Ilyitch." To clinical physicians especially, the story is terrifying in its uncanny accuracy and in the lessons it teaches. Tolstoy wrote as though possessed of an inborn knowledge greater than any he could possibly have acquired in life. How else could he have intuited the terrible solitude of a death made lonely by withholding the truth, "this solitude through which he [Ivan Ilyitch] was passing, as he lay with his face turned to the back of the divan,—a solitude amid a populous city, and amid his numerous circle of friends and family,—a solitude deeper than which could not be found anywhere, either in the depths of the sea, or in the earth . . ."? Ivan could share his terrible knowledge with no one, "and he had to live thus on the edge of destruction—alone, without anyone to understand and pity him."

Ivan was not surrounded by people who loved him, and in part perhaps this was why he resorted to wishing, at least a little, to be the object of pity, a graceless state to which few of us would willingly fall at the end of life. The origin of his wife's attempted deception seems to have been her own determination not to deal with the emotional consequences that the truth would precipitate. Whether such deceptions arise from scorn or from misguided affection, they always leave their victim to deal with his leave-taking alone. In her case, a patronizing contempt was the basis on which she convinced herself that her husband's death would be easier for both of them if it went undiscussed. It was herself she was thinking about, and not her husband, whose mortal illness was an inconvenience to her, and even an imposition on her household. In this atmosphere, Ivan could not find the strength to confront the result had he forced the issue:

> Ivan Ilyitch's chief torment was a lie,—the lie somehow accepted by everyone, that he was only sick, but not dying, and that he needed only to be calm, and trust to the doctors, and then somehow he would come out all right. But he knew that, whatever was done, nothing would come of it, except still more excruciating anguish and death. And this lie tormented him; it tormented him that they were unwilling to acknowledge what all

> knew as well as he knew, but preferred to lie to him about his terrible situation, and made him also a party to the lie. This lie, this lie, it clung to him, even to the very evening of his death; this lie, tending to reduce the strange, solemn act of his death to the same level as visits, curtains, sturgeon for dinner—it was horribly painful for Ivan Ilyitch. And strange! many times, when they were playing this farce for his benefit, he was within a hair's breadth of shouting at them:
>
> "Stop your foolish lies! you know as well as I know that I am dying, and so at least stop lying."
>
> But he never had the spirit to do this.

There is another element, too, that these days often conspires to isolate the mortally ill. I can think of no better word for it than *futility*. Pursuing treatment against great odds may seem like a heroic act to some, but too commonly it is a form of unwilling disservice to patients; it blurs the borders of candor and reveals a fundamental schism between the best interests of patients and their families on the one hand and of physicians on the other.

The Hippocratic philosophy of medicine declares that nothing should be more important to a physician than the best interests of the patient who comes to him for care. Although we live now in an era when the needs of the greater society sometimes come into conflict with a doctor's judgment concerning what is best for his individual patient, there has never been any doubt that the goal of medical care is to overcome sickness and relieve suffering. Every medical student learns very early that it is sometimes necessary to add for a time to a patient's suffering in order to overcome his sickness, and there are few people who do not understand and accept that necessity. This is especially true for the hundred or more diseases that comprise the various forms of cancer, where combinations of effective surgery, radiation, and chemotherapy commonly result in periods of debility and other severe temporary torments, if not frank complications. Few people faced with a diagnosis of potentially remediable malignant disease should be willing to give up the struggle if there is any reasonable chance that some promising form of treatment is available to lessen the ravages of the disease or cure it. To do anything less is not stoicism, but folly.

Once more, the dilemma faced by all of us when we find ourselves in these situations lies in the use of language. Here, the operative obscurities are words such as *reasonable* and *promising.* It is in such seemingly clear but actually ambiguous terminology that clues appear, exposing the schism often existing between the goals of doctors and the goals of the people they treat. At the cost of burdening these pages with even more autobiography, I shall use my own professional evolution as a physician to illustrate the subtle progression by which a young medical student who wants only to care for his sick fellows becomes transmuted unawares into the embodiment of a biomedical problem-solver.

Before there were two digits in my age, I had seen the hope (I choose the word deliberately) that a doctor's presence brings to a worried family. There were several frightening emergencies during my mother's long illness, even in the years before she had begun her descent to death. The mere knowledge that someone had gone to the drugstore phone to call the doctor, and the word that he was on the way, changed the atmosphere in our small apartment from terrified helplessness to a secure sense that somehow the dreadful situation could be made right. That man—the man who stepped across the threshold with a smile and an air of competence, who called each of us by name, who understood that beyond anything else we needed reassurance, and whose very entrance into our home conveyed it—that was the man I wanted to be.

My objective in becoming a physician was to be a general practitioner in the Bronx. In the first year of medical school, I learned how the body functions; in the second year, I learned how it gets sick. In the third and fourth years, I began to understand how to interpret the histories I elicited from my patients and to study the physical and chemical clues produced by their illnesses, that combination of overt and hidden findings that the eighteenth-century pathologist Giovanni Morgagni called "the cries of the suffering organs." I studied the various ways of listening to my patients and looking at them so that I might be able to discern those cries. I was taught to probe orifices, read X rays, and seek meaning in the state of blood and cast-off waste products of various descriptions. In time, I knew exactly which tests to order so that the more obvious clues might be used to lead me to the hidden changes that are part of sickness. That process is pathophysiology. Mastering its tortuous patterns is the means by which to understand the details of the way normal mechanisms of healthy life somehow go awry. To understand pathophysiology is to hold the key to diagnosis, without which there can be no cure. The quest of every doctor in approaching serious disease is to make the diagnosis and design and carry out the specific cure. This quest, I call The Riddle, and I capitalize it so there will be no mistaking its dominance over every other consideration. The satisfaction of solving The Riddle is its own reward, and the fuel that drives the clinical engines of medicine's most highly trained specialists. It is every doctor's measure of his own abilities; it is the most important ingredient in his professional self-image.

By the time I finished medical school, I had discovered greater dimensions in the pursuit of diagnosis and ever-expanding challenges in carrying out successful treatment. The goal became to understand the evolution of a disease process so well that it could be combatted with exactly the right choices of excision, repair, biochemical modification, or any of the increasing variety of modalities that constantly make their appearance. The six years of my residency training was preparation for dealing with each aspect of The Riddle, which by the end of that time had become the fascination of my life. In me, my teachers had replicated themselves.

I had given up any thought of returning to be a local doctor in the Bronx or any place like it. I never forgot the need to be to my patients what the general practitioner had been to our family, but I realize now that his image was no longer the one I most admired. I was totally absorbed with The Riddle, and the doctor who inspired me was the doctor who was best at solving it.

All of my professional life, I have tried, as I believe the great majority of physicians do, to be the kind of doctor whose example led me to choose healing as my life's work. But alongside that example has been another, more powerful image—the challenge that motivates most persuasively; the challenge that makes each of us physicians continue ever trying to improve our skills; the challenge that results in the dogged pursuit of a diagnosis and a cure; the challenge that has resulted in the astounding progress of late-twentieth-century clinical medicine—that foremost of challenges is not primarily the welfare of the individual human being but, rather, the solution of The Riddle of his disease.

We seek to treat our patients with the empathy that is so major a factor in their recovery, and we always try to guide them in making decisions that we think will lead to relief of their suffering. But that is not enough to sustain and improve our abilities, or even to maintain our enthusiasm. It is The Riddle that drives our most highly skilled and the most dedicated of our physicians.

In one of his *Precepts*, Hippocrates wrote, "Where love of mankind is, there is also love of the art of medicine," and that is as true as it has ever been; were it otherwise, the burden of caring for our fellows would soon prove unbearable. Nevertheless, our most rewarding moments of healing derive not from the works of our hearts but from those of our intellects—it is there that the passion is most intense. I have come to realize the truth, and even the necessity that it should be so. As doctors, we must confront that about ourselves every time we undertake to care for another human being; as patients, we must understand that a physician's driving quest to solve The Riddle will sometimes be at odds with our best interests at the end of life.

Every medical specialist must admit that he has at times convinced patients to undergo diagnostic or therapeutic measures at a point in illness so far beyond reason that The Riddle might better have remained unsolved. Too often near the end, were the doctor able to see deeply within himself, he might recognize that his decisions and advice are motivated by his inability to give up The Riddle and admit defeat as long as there is any chance of solving it, Though he be kind and considerate of the patient he treats, he allows himself to push his kindness aside because the seduction of The Riddle is so strong and the failure to solve it renders him so weak.

Patients are awed by their doctors, create a transference with them in the true psychoanalytic sense, and wish to please them, or at least not to be seen as a source of offense. Some believe that doctors always know exactly what they are doing, and that uncertainty is utterly alien to the superspecialists who treat the most seriously ill people in the hospital. They are convinced—and the more high-tech the doctor, the more their patients are convinced—that the men and women who treat them always have very good scientific reasons for recommending the courses of action they do.

Patients often have substantial reasons for not going further when only a diminishingly small possibility exists that they may survive. Some reasons are philosophical or spiritual, some are quite practical, and some arise simply from the conviction that what one gets after a major struggle for recovery is not worth what has to be endured in order to get it. As a very wise oncology nurse once told me, "For some people, even the certainty of coming out on the other side of weeks of distress doesn't justify the physical and emotional price they have to pay."

Beside me as I write these paragraphs lies the chart of Miss Hazel Welch, a ninety-two-year-old woman who lived in the convalescent unit of a senior citizen's residence complex about five miles from the Yale–New Haven Hospital. Although mentally alert, she required the nursing care of the unit because of such advanced arthritis and arteriosclerotic obstruction in the arteries of her legs that she could no longer walk unassisted; at the time of the acute illness for which I treated her, she was on the semielective list for amputation of one of the toes of her left foot, which had become gangrenous. She was taking antiinflammatory medications for severe arthritis and was in remission from chronic leukemia. "Here a pivot, there a wheel, now a pinion, next a spring" were giving way, and Jefferson might have counseled me that it was folly to attempt to prevent the whole machine from surceasing motion entirely.

Shortly after noon on February 23, 1978, Miss Welch fell to the floor unconscious in the presence of one of the nursing aides. An ambulance took her to the emergency room of the Yale–New Haven Hospital, where she was found to have no measurable blood pressure; the physical findings were consistent with severe peritonitis. After the rapid intravenous infusion of fluids, she was sufficiently resuscitated to undergo a quick X-ray examination, which revealed a large amount of air free in her abdominal cavity. The diagnosis was clear: She undoubtedly had a perforated digestive tract, and the most likely source was an ulcer of the duodenum, just beyond the stomach.

By then completely alert and rational, Miss Welch refused an operation. In a broad Yankee inflection, she told me that she had been on this planet "quite long enough, young man" and didn't wish to go on. There was no one, she said, to live for—the space on her chart's top page for next of kin bore the name of a trust officer at the Connecticut National Bank. To me, standing at the side of her gurney perfectly healthy and embosomed outside of that place by family and friends, her decision made no sense. I used every argument I could muster in trying to persuade her that the crystal clarity of her brain and the responsiveness of her leukemia meant that she had good years ahead. I was completely frank in telling her that, given the state of her atherosclerosis and the peritonitis, her chance of recovery from the required surgery was only about one in three. "But," I said, "one in three, Miss Welch, is a lot better than certain death, which is what happens if you don't let us operate." That seemed self-evident, and I couldn't imagine that anyone as obviously sensible as she could possibly believe otherwise. She remained adamant, and I left her alone to think about it, her chances of survival decreasing as the minutes ticked by.

I returned a quarter of an hour later. My patient was positioned half-upright on the gurney, scowling at me as though I were a middle-aged naughty boy. She reached out and took my hand, staring hard directly into my eyes as though charging me with a grave mission for whose failure she would hold me personally responsible. "I'll do it," she said, "but only because I trust you." Suddenly, I felt a little less sure I was doing the right thing.

During the operation, I discovered a duodenal perforation so massive that its repair required much more extensive surgery than I had anticipated. The stomach had become almost completely separated from the duodenum, as though exploded away from it; Miss Welch's abdomen was filled with corrosive digestive juices and whole pieces of the lunch she had eaten a few minutes before collapsing. I did what was necessary, closed the abdomen, and admitted my still-unconscious patient to the surgical intensive care unit. She had inadequate respiratory drive to breathe, so the anesthesiologist's tube remained in her windpipe.

At the end of a week, Miss Welch was improving, although she was not mentally alert enough to understand what was happening around her. Finally, her mind cleared completely, and until the breathing tube could be removed from between her vocal cords two days later, she spent every minute of my twice-a-day-visits staring reproachfully at me. When she was able to speak, she lost no time in letting me know what a dirty trick I had pulled by not letting her die as she wished. I indulged her in this, certain that I had done the right thing, and with living evidence, I thought, to prove it. She had, after all, survived. But she saw things differently and didn't hesitate to let me know I had betrayed her by minimizing the difficulties of the postoperative period. Knowing Miss Welch would have refused life-saving surgery if she had been aware of the kinds of things elderly arteriosclerotic people often endure in surgical intensive care units, I had in my description of the anticipated postoperative days played down what she could realistically be expected to experience. She had been through too much, she said, and she didn't trust me anymore. She was obviously one of those people to whom survival was not worth the cost, and I had not been completely forthcoming in predicting what that cost might be. Although my intentions were only to serve what I conceived to be her welfare, I was guilty of the worst sort of paternalism. I had withheld information because I was afraid the patient might use it to make what I thought of as a wrong decision.

Two weeks after her transfer back to her old room at the residence unit, Miss Welch had a massive stroke and died in less than a day. In keeping with the instructions she had written in the presence of her trust officer on his first visit after her hospital discharge, no attempt was made to give her anything but nursing care. She wanted no repetition of her recent experience and emphatically said so in her written statement. Although the trauma of her peritonitis and the surgery had obviously strengthened the likelihood of her stroke, I suspect that her continuing anger at my well-intentioned deception also played a role. But perhaps the most important factor in my patient's death may simply have been her wish not to continue living, which had been frustrated by my ill-advised operation. I had won out over The Riddle but lost the greater battle of humane patient care.

Had I carefully considered the factors I have described in this book's chapters on aging, I would not have been so quick to recommend an operation. For Miss Welch, the effort was not justified, no matter what success might have resulted, and I was not wise enough to recognize it. I see things differently now. Had I the chance to relive this episode, or some others like it in my career, I would listen more to the patient and ask her less to listen to me. My objective was to grapple with The Riddle; hers was to use this sudden illness as a gracious way to die. She gave in only to please me.

There is a lie in the paragraph you have just read. I imply there that I would have acted differently, although I know I would probably have done exactly the same thing again, or risk the scorn of my peers. It is in such matters that ethicists and moralists run aground when they try to judge the actions of bedside doctors, because they cannot see the trenches from their own distant viewing point. The code of the profession of surgery demands that no patient as salvageable as Miss Welch be allowed to die if a straightforward operation can save her, and we who would break that fundamental rule, no matter the humaneness of our motive, do so at our own peril. Viewed by a surgeon, mine was strictly a clinical decision, and ethics should not have been a consideration. Had I let Miss Welch have her way, I would have had to defend the result at the weekly surgical conference (where it would certainly be seen as *my* decision, not hers), before unbending colleagues to whom her death would seem a case of poor judgment, if not downright negligence of the clear duty to save life. I would almost certainly be castigated over my failure to overrule such a seemingly senseless wish. I can imagine what I might hear: "How could you let her talk you into it?" "Does the mere fact that an old lady wants to die mean you should be a party to it?" "A surgeon should only make clinical decisions, and the right clinical decision was to operate—leave moralizing to the ministers!" This is a form of peer pressure to which I will not be presumptuous enough to claim immunity. One way or another, the rescue credo of high-tech medicine wins out, as it almost always does.

My treatment of Miss Welch was based not on her goals but on mine, and on the accepted code of my specialty. I pursued a form of futility that deprived her of the particular kind of hope she had longed for—the hope that she could leave this world without interference when an opportunity arose. No matter her lack of family, the nurses and I could have seen to it that she did not die alone, at least insofar as empathetic strangers can do this for a friendless old person. Instead, she suffered the fate of so many of today's hospitalized dying, which is to be separated from reality by the very biotechnology and professional standards that are meant to return people to a meaningful life.

The beeping and squealing monitors, the hissings of respirators and pistoned mattresses, the flashing multicolored electronic signals—the whole technological panoply is background for the tactics by which we are deprived of the tranquillity we have every right to hope for, and separated from those few who would not let us die alone. By such means, biotechnology created to provide hope serves actually to take it away, and to leave our survivors bereft of the unshattered final memories that rightly belong to those who sit nearby as our days draw to a close.

Every scientific or clinical advance carries with it a cultural implication, and often a symbolic one. The invention of the stethoscope in 1816, for example, can be viewed as having set in motion the process by which physicians came to distance themselves from their patients. Such an interpretation of the instrument's role was, in fact, considered by some medical commentators of the time to be one of its advantages, since not many clinicians, then or now, feel at ease with an ear pressed up against a diseased chest. That and its image as a visible evidence of status remain to this day unspoken reasons for the instrument's popularity. One need only spend a few hours on rounds with young resident physicians to observe the several roles played by this dangling evidence of authority and detachment.

Seen from the strictly clinical perspective, a stethoscope is nothing more than a device to transmit sounds; by the same kind of reasoning, an intensive care unit is merely a secluded treasure room of high-tech hope within the citadel in which we segregate the sick so that we may better care for them. Those tucked-away sanctums symbolize the purest form of our society's denial of the naturalness, and even the necessity, of death. For many of the dying, intensive care, with its isolation among strangers, extinguishes their hope of not being abandoned in the last hours. In fact, they *are* abandoned, to the good intentions of highly skilled professional personnel who barely know them.

Nowadays, the style is to hide death from view. In his classic exposition of the customs associated with dying, the French social historian Philippe Ariès calls this modern phenomenon the "Invisible Death." Dying is ugly and dirty, he points out, and we do not easily tolerate anymore what is ugly and dirty. Death is therefore to be secluded and to occur in sequestered places:

> The hidden death in the hospital began very discreetly in the 1930's and 1940's and became widespread after 1950. . . . Our senses can no longer tolerate the sights and smells that in the early nineteenth century were part of daily life, along with suffering and illness. The physiological effects have passed from daily life to the aseptic world of hygiene, medicine and morality. The perfect manifestation of this world is the hospital, with its cellular discipline. . . . Although it is not always admitted, the hospital has offered families a place where they can hide the unseemly invalid whom neither the world nor they can endure. . . . The hospital has become the place of solitary death.

Eighty percent of American deaths now occur in the hospital. The figure has gradually risen since 1949, when it was 50 percent; in 1958, it reached 61 percent, and in 1977, it was 70 percent. The increase is not only because so many of the dying have needed the high level of acute care that can be provided only within the hospital's walls. The cultural symbolism of sequestering the dying is here as meaningful as the strictly clinical perspective of improved access to specialized facilities and personnel, and for most patients even more so.

The solitary death is now so well recognized that our society has organized against it, and well we should. From the wisdom of the legal documents called advance directives to the questionable philosophies of suicide societies, a range of options exists, and at bottom the goal of each of them is the same: a restoration of certainty that when the end is near, there will be at least this source of hope—that our last moments will be guided not by the bioengineers but by those who know who we are.

This hope, the assurance that there will be no unreasonable efforts, is an affirmation that the dignity to be sought in death is the appreciation by others of what one has been in life. It is a dignity that proceeds from a life well lived and from the acceptance of one's own death as a necessary process of nature that permits our species to continue in the form of our own children and the children of others. It is also the recognition that the *real* event taking place at the end of our life is our death, not the attempts to prevent it. We have somehow been so taken up with the wonders of modern science that our society puts the emphasis in the wrong place. It is the dying that is the important thing—the central player in the drama is the dying man; the dashing leader of that bustling squad of his would-be rescuers is only a spectator, and a groundling at that.

In ages past, the hour of death was, insofar as circumstances permitted, seen as a time of spiritual sanctity, and of a last communion with those being left behind. The dying expected this to be so, and it was not easily denied them. It was their consolation and the consolation of their loved ones for the parting and especially for the miseries that had very likely preceded it. For many, this last communion was the focus not only of the sense that a good death was being granted them but of the hope they saw in the existence of God and an afterlife.

It is ironic that in redefining hope, I should find it necessary to call attention to what was until recently the very precinct in which most people would seek it. Much less commonly than at any other time in this millennium do the dying nowadays turn to God and the promise of an afterlife when the present life is fading. It is not for medical personnel or skeptics to question the faith of another, particularly when that other is facing eternity. Agnostics and even atheists have been known to find solace in religion at such times, and their drastic changes of heart are to be respected. How many times, when I was a young surgeon, did I hear a physician or nurse scoff at the sacrament of extreme unction because "It's just like telling him that he's about to die," and then see him or her slow to call the priest whose presence, had the patient only known the truth, he would have preferred over the doctor's? Years ago my hospital had an illness category called the Danger List. When a Catholic's name was entered on it, his priest was automatically summoned. Among the several reasons such a list is no longer in existence is the official reluctance to "scare" a patient by the appearance in his room of someone with a clerical collar, because this has been so often a person's first intimation that his life is waning. In such ways did hospital officialdom deny hope, and even religious faith was subverted to accomplish it.

Sometimes a dying person's source of hope can be as undemanding as the wish to live until a daughter's graduation or even a holiday that has particular meaning. The medical literature documents the power of this kind of hope, describing instances in which it has maintained not only the life but the optimism of a dying man or woman for the necessary period. Every doctor and many laymen can tell of individuals who survived weeks beyond the most extreme expectations in order to have one last Christmas or to await the sight of a dear face arriving from some distant land.

The lesson in all of this is well known. Hope lies not only in an expectation of cure or even of the remission of present distress. For dying patients, the hope of cure will always be shown to be ultimately false, and even the hope of relief too often turns to ashes. When my time comes, I will seek hope in the knowledge that insofar as possible I will not be allowed to suffer or be subjected to needless attempts to maintain life; I will seek it in the certainty that I will not be abandoned to die alone; I am seeking it now, in the way I try to live my life, so that those who value what I am will have profited by my time on earth and be left with comforting recollections of what we have meant to one another.

There are those who will find hope in faith and their belief in an afterlife; some will look forward to the moment a milestone is reached or a deed is accomplished; there are even some whose hope is centered on maintaining the kind of control that will permit them the means to decide the moment of their death, or actually to make their own quietus unhindered. Whatever form it may take, each of us must find hope in his or her own way.

There is a specific form of abandonment that is particularly common among patients near death from cancer, and it requires comment. I refer here to abandonment by doctors. Doctors rarely *want* to give up. As long as there is any possibility of solving The Riddle, they will keep at it, and sometimes it takes the intervention of a family or the patient himself to put an end to medical exercises in futility. When it becomes obvious, though, that there is no longer a Riddle on which to focus, many doctors lose the drive that sustained their enthusiasm. As the long siege drags on and one after another treatment has begun to fail, those enthusiasms tend to fall by the wayside. Emotionally, doctors then tend to disappear; physically, too, they sometimes all but disappear.

Many reasons have been cited to explain why physicians abandon patients when they are beyond recovery. Studies are pointed to, indicating that of all the professions, medicine is the one most likely to attract people with high personal anxieties about dying. We become doctors because our ability to cure gives us power over the death of which we are so afraid, and loss of that power poses such a significant threat that we must turn away from it, and therefore from the patient who personifies our weakness. Doctors are people who succeed—that is how they survived the fierce competition to achieve their medical degree, their training, and their position. Like other highly talented people, they require constant reassurance of their abilities. To be unsuccessful is to endure a blow to self-image that is poorly tolerated by members of this most egocentric of professions.

I have also been impressed with another factor in the personalities of many doctors, perhaps linked to the fear of failure: a need to control that exceeds in magnitude what most people would find reasonable. When control is lost, he who requires it is also a bit lost and so deals badly with the consequences of his impotence. In an attempt to maintain control, a doctor, usually without being aware of it, convinces himself that he knows better than the patient what course is proper. He dispenses only as much information as he deems fit, thereby influencing a patient's decision-making in ways he does not recognize as self-serving. This kind of paternalism was precisely the source of my error in treating Miss Welch.

The inability to face the consequences presented by loss of control often leads a physician to walk away from situations in which his power no longer exists, and this must certainly be an ingredient in the abrogation of responsibility that so often takes place at the end of a patient's life. In the structured formulation he sees in The Riddle and in the systematic way he goes about its solution, the doctor creates order from chaos and finds the power to exert control over disease, nature, and his personal universe. When there is no longer a Riddle, such a doctor will lower his interest or lose it entirely. To stay and oversee the triumph of unrestrainable nature is to acquiesce to his own impotence.

Or, having lost the major battle, the doctor may maintain a bit of authority by exerting his influence over the dying process, which he does by controlling its duration and determining the moment at which he allows it to end. In this way, he deprives the patient and family of the control that is rightfully theirs. These days, many hospitalized patients die only when a doctor has decided that the right time has come. Beyond the curiosity and the problem-solving challenge fundamental to good research, I believe that the fantasy of controlling nature lies at the very basis of modern science. Even with all its art and philosophy, the modern profession of medicine has become, to a great extent, an exercise in applied science, with the goal of that conquest in mind. The ultimate aim of the scientist is not only knowledge for the sake of knowledge, but knowledge with the aim of overcoming that in our environment which he views as hostile. None of the acts of nature (or Nature) is more hostile than death. Every time a patient dies, his doctor is reminded that his own and mankind's control over natural forces is limited and will always remain so. Nature will always win in the end, as it must if our species is to survive.

The necessity of nature's final victory was expected and accepted in generations before our own. Doctors were far more willing to recognize the signs of defeat and far less arrogant about denying them. Medicine's humility in the face of nature's power has been lost, and with it has gone some of the moral authority of times past. With the vast increase in scientific knowledge has come a vast decrease in the acknowledgment that we still have control over far less than we would like. Physicians accept the conceit (in every sense of the word) that science has made us all-powerful and therefore the only proper judges of how our skills are to be used. The greater humility that should have come with greater knowledge is instead replaced by medical hubris: Since we can do so much, there is no limit to what should be attempted—*today*, and for *this patient!*

The more highly specialized the physician, the more likely is The Riddle to be his primary motivation. To medicine's absorption with The Riddle, we owe the great clinical advances of which all patients are the beneficiaries; to medicine's absorption with

The Riddle, we also owe our disappointment when we cherish expectations of doctors that they cannot fulfill and perhaps should not be asked to fulfill. The Riddle is the doctor's lodestone as an applied scientist; it is his albatross as a humane caregiver.

Oncologists are among the most determined of medical people, prepared to try almost any last-ditch effort to stave off inevitability—they can be seen on the barricades when other defenders have furled their flags. Like so many of their specialized colleagues, oncologists can be empathetic and beneficent; when they deal with patients, they are likely to review treatment and complications at length, lay out courses of action, and develop warm relationships with individuals and families alike. And yet they so often do it without ever being able to come to a real understanding of the spiritual nature of those they treat or of their subjective response to the looming face of death that always oversees their efforts. Sad to say, this is true of the great majority of the specialists who treat our most complex diseases. As I look back on my thirty years of practice, I am increasingly made aware that I have been much more the problem-solver than the man in the Bronx whose only wish was to nurture his patients.

If we should no longer expect from so many of our doctors what they cannot give, how are we, as patients, to be guided in making rational decisions? In the first place, those doctors can still guide us. In fact, the information they impart becomes even more valuable once we adjust to using it only as a way of comprehending the pathophysiology they know so well. Knowing that they are without the power to dominate our judgment, our specialists will be less prone to tell us things in a way that influences the decision they want us to make. It behooves every patient to study his or her own disease and learn enough about it to recognize the onset of that time when further treatment becomes a debatable issue. Such an education begins with learning how the normal body works, which much simplifies familiarity with the ways in which it is affected by disease. Clearly, cancer is a process particularly well suited to such an approach, and it should not be beyond the capacities of any but a small percentage of people to accomplish it.

In discussing The Riddle, I have not written about the sort of doctor who is much less under its spell than is the specialist. The relationship between a patient and his primary doctor will remain the core of cure, as it has been since the days when Hippocrates set down his reflections upon it. When there can be no cure, that relationship takes on an importance of immeasurable magnitude.

It would behoove our government to support the concept of family practice and the primary care that should be the major focus of any scheme of health delivery. Funding for its training programs in medical schools and teaching hospitals deserves to become a major priority, and the dedication of talented young people should be encouraged. Of all the possible advantages of such a system, I can think of none more valuable than the humanizing effect it would have on the way we die. So much must be borne at the time of death we should not add to it by asking advice only from specialized strangers, when it is possible to be guided with the insight of a long-standing relationship with our own doctor.

We bear more than pain and sorrow when we depart life. Among the heaviest burdens is apt to be regret, which deserves a word at this point. As inevitable as death is and as likely to be preceded by a difficult period, especially for people with cancer, there are additional pieces of baggage we shall all take to the grave, but from which we may somewhat disencumber ourselves if we anticipate them. By these, I mean conflicts unresolved, breached relationships not healed, potential unfulfilled, promises not kept, and years that will never be lived. For virtually every one of us, there will be unfinished business. Only the very old escape it, and even then not always.

Perhaps the mere existence of things undone should be a sort of satisfaction in itself, though the idea would appear to be paradoxical. Only one who is long since dead while still seemingly alive does not have many "promises to keep, and miles to go before I sleep," and that state of inertness is not to be desired. To the wise advice that we live every day as though it will be our last, we do well to add the admonition to live every day as though we will be on this earth forever.

We do well also to avoid another unnecessary burden by remembering the caution of Robert Burns about the best-laid plans. Death rarely, if ever, acts according to our plans or even to our expectations. Everyone wants to do this thing of dying in the proper way, a modern version of *ars moriendi* and the beauty of final moments. Since human beings first began to write, they have recorded their wish for an idealized ending some call the "good death," as if any of us can ever be sure of it or have any reason to expect it. There are pitfalls of decision-making to be sidestepped and varieties of hope to seek, but beyond that we must forgive ourselves when we cannot achieve some preconceived image of dying right.

Nature has a job to do. It does its job by the method that seems most suited to each individual whom its powers have created. It has made this one susceptible to heart disease and that one to stroke and yet another to cancer, some after a long time on this earth and some after a time much too brief, at least by our own reckoning. The animal economy has formed the circumstances by which each generation is to be succeeded by the next. Against the relentless forces and cycles of nature there can be no lasting victory.

When at last the moment comes and the perception is inescapable that we have reached the point where, with Browning's Jochanan Hakkadosh, our "feet tread the way of all flesh," it is incumbent on us to remember that it is not only the way of all flesh but the way of all life, and it has its own plans for us. Though we find clever ways to delay, there is no way to undo those plans. Even suicides yield to the cycle, and for all we know the stimulus for the action they take has been designed in some vast scheme that is just another example of the immutable laws of nature and its animal economy. Shakespeare has Julius Caesar reflect that:

> Of all the wonders that I yet have heard,
> It seems to me most strange that men should fear;
> Seeing that death, a necessary end,
> Will come when it will come.

Epilogue

I AM MORE concerned with the microcosm than the macrocosm; I am more interested in how a man lives than how a star dies; how a woman makes her way in the world than how a comet streaks across the heavens. If there is a God, He is present as much in the creation of each of us as He was at the creation of the earth. The human condition is the mystery that engages my fascination, not the condition of the cosmos.

To understand the human condition has been the work of my life. During that life, which is now into its seventh decade, I have had my share of sorrows and my share of triumphs. Sometimes I think I have had far more than my share of both, but that impression probably stems from the inclination we all share, which makes each of us see our own existence as a heightened example of universal experience—a life that is somehow larger than life, and felt more deeply.

There is no way to foretell whether this is to be my last decade or whether there will be more—good health is a guarantee of nothing. The only certainty I have about my own death is another of those wishes we all have in common: I want it to be without suffering. There are those who wish to die quickly, perhaps with instantaneous suddenness; there are those who wish to die at the end of a brief, anguish-free illness, surrounded by the people and the things they love. I am one of the latter, and I suspect I am in the majority.

What I hope, unfortunately, is not what I expect. I have seen too much of death to ignore the overwhelming odds that it will not occur as I wish it. Like most people, I will probably suffer with the physical and emotional distress that accompany many mortal illnesses, and like most people I will probably compound the pained uncertainty of my last months by the further agony of indecision—to continue or to give in, to be treated aggressively or to be comforted, to struggle for the possibility of more time or to call it a day and a life—these are the two sides of the mirror into which we look when afflicted by those illnesses that have the power to kill. The side in which we choose to see ourselves reflected during the last days should reveal an image that is tranquil in its decision, but even that is not to be counted on.

I have written this book as much for myself as for everyone who reads it. By trooping some of the army of the horsemen of death across the field of our vision, I hope to recall the things I have seen, and make them familiar to everyone else. There is no need to look at the whole long line of mounted murderers; there are far more cavalrymen than any of us could have the stomach for. But they all use weapons not much different from those you have been reading about.

If they become just a bit more familiar, perhaps these horsemen will also become less frightening, and perhaps those decisions that must be made can be sought out in an atmosphere less charged with half-knowledge, anxiety, and unjustified expectations. For each of us, there may be a death that is the right death, and we should strive to find it, while accepting that it may prove ultimately to be beyond our grasp. The final disease that nature inflicts on us will determine the atmosphere in which we take our leave of life, but our own choices should be allowed, insofar as possible, to be the decisive factor in the manner of our going. Rilke wrote:

> Oh Lord, give each of us his own death,
> The dying, that issues forth out of the life
> In which he had love, meaning and despair.

The poet states it as a prayer, and, like all prayers, it may not be possible to answer it, even for God. For too many of us, the manner of death will prove to be beyond control, and no knowledge or wisdom can change that. During the dying of someone we love or of ourselves, it is valuable to know that there are still far too many things about which circumstance does not allow choices, even with the finest and most beneficently motivated of the forces of modern biomedical science on our side. It is not a judgment upon the many that they are fated to die badly, simply the nature of the thing that kills them.

The great majority of people do not leave life in a way they would choose. In previous centuries, men believed in the concept of *ars moriendi*, the art of dying. Those were times when the only possible attitude to the approach of death was to let it happen—once certain symptoms made their appearance, there was no choice but to die the best way possible, at peace with God. But even then, most people went through a period of suffering that preceded the end; there was little but resignation and the consolation of prayer and family to ease the final time.

We live today in the era not of the art of dying, but of the art of saving life, and the dilemmas in that art are multitudinous. As recently as half a century ago, that other great art, the art of medicine, still prided itself on its ability to manage the process of death, making it as tranquil as professional kindness could. Except in the too-few programs such as hospice, that part of the art is now mostly lost, replaced by the brilliance of rescue and, unfortunately, the all-too-common abandonment when rescue proves impossible.

Death belongs to the dying and to those who love them. Though it may be sullied by the incursive havoc of disease, it must not be permitted to be further disrupted by well-meant exercises in futility. Decisions about continuation of treatment are influenced by the enthusiasm of the doctors who propose them. Commonly, the most accomplished of the specialists are also the most convinced and unyielding believers in biomedicine's ability to overcome the challenge presented by a pathological process close to claiming its victim. A family grasps at a straw that comes in the form of a statistic; what is offered as objective clinical reality is often the subjectivity of a devout disciple of the philosophy that death is an implacable enemy. To such warriors, even a temporary victory justifies the laying waste of the fields in which a dying man has cultivated his life.

I say these things not to condemn high-tech doctors. I have been one of them, and I have shared the excitement of last-ditch fights for life and the supreme satisfaction that comes when they are won. But more than a few of my victories have been Pyrrhic. The suffering was sometimes not worth the success. I also believe that had I been able to project myself into the place of the family and the patient, I would have been less often certain that the desperate struggle should be undertaken.

When I have a major illness requiring highly specialized treatment, I will seek out a doctor skilled in its provision. But I will not expect of him that he understand my values, my expectations for myself and those I love, my spiritual nature, or my philosophy of life. That is not what he is trained for and that is not what he will be good at. It is not what drives those engines of his excellence.

For those reasons, I will not allow a specialist to decide when to let go. I will choose my own way, or at least make the elements of my own way so clear that the choice, should I be unable, can be made by those who know me best. The conditions of my illness may not permit me to "die well" or with any of the dignity we so optimistically seek, but within the limits of my ability to control, I will not die later than I should simply for the senseless reason that a highly skilled technological physician does not understand who I am.

Between the lines of this book lies an unspoken plea for the resurrection of the family doctor. Each one of us needs a guide who knows *us* as well as he knows the pathways by which we can approach death. There are so many ways to travel through the same thickets of disease, so many choices to make, so many stations at which we may choose to rest, continue, or end the journey completely—until the last steps of that journey we need the company of those we love, and we need the wisdom to choose the way that is ours alone. The clinical objectivity that should enter into our decisions must come from a doctor familiar with our values and the lives we have led, and not just from the virtual stranger whose superspecialized biomedical skills we have called upon. At such times, it is not the kindness of strangers we need, but the understanding of a longtime medical friend. In whatever way our system of health care is reorganized, good judgment demands that this simple truth be appreciated.

And yet, even with the most sensitive medical ombudsman, real control requires one's own knowledge about the ways of sickness and death. Just as I have seen people struggle too long, I have also seen others give up too early, when there is still much that can be done to preserve not only life but enjoyment as well. The more knowledge we have about the realities of lethal illness, the more sensible we can be about choosing the time to stop or the time to fight on, and the less we expect the kind of death most of us will not have. For those who die and those who love them, a realistic expectation is the surest path to tranquillity. When we mourn, it should be the loss of love that makes us grieve, not the guilt that we did something wrong.

A realistic expectation also demands our acceptance that one's allotted time on earth must be limited to an allowance consistent with the continuity of the existence of our species. Mankind, for all its unique gifts, is just as much a part of the ecosystem as is any other zoologic or botanical form, and nature does not distinguish. We die so that the world may continue to live. We have been given the miracle of life because trillions upon trillions of living things have prepared the way for us and then have died—in a sense, for us. We die, in turn, so that others may live. The tragedy of a single individual becomes, in the balance of natural things, the triumph of ongoing life.

All of this makes more precious each hour of those we have been given; it demands that life must be useful and rewarding. If by our work and pleasure, our triumphs and our failures, each of us is contributing to an evolving process of continuity not only of our species but of the entire balance of nature, the dignity we create in the time allotted to us becomes a continuum with the dignity we achieve by the altruism of accepting the necessity of death.

How important, then, is the actual deathbed scene of serenity and leave-taking? For most of us, it will prove to be an image of wishfulness, an ideal to be striven for, perhaps even approached, but usually not attained except by comparatively few, whose circumstances of terminal illness permit it to come to pass.

The rest of us must make do with what we will be given. Through an understanding of the mechanisms by which the commonest mortal diseases kill, through the wisdom that comes of realistic expectations, through a new understanding with our physicians that we do not ask of them what they cannot give, the end can be managed with the greatest degree of control allowed by the pathological process that kills us.

Though the hour of death itself is commonly tranquil and often preceded by blissful unawareness, the serenity is usually bought at a fearful price—and the price is the process by which we reach that point. There are some who manage to achieve moments of nobility in which they somehow transcend the indignities being visited on them, and these moments are to be cherished. But such intervals do not lessen the distress over which they briefly triumph. Life is dappled with periods of pain, and for some of us is suffused with it. In the course of ordinary living, the pain is mitigated by periods of peace and times of joy. In dying, however, there is only the affliction. Its brief respites and ebbs are known always to be fleeting and soon succeeded by a recurrence of the travail. The peace, and sometimes the joy, that may come occurs with the release. In this sense, there is often a serenity—sometimes even a dignity—in the act of death, but rarely in the process of dying.

And so, if the classic image of dying with dignity must be modified or even discarded, what is to be salvaged of our hope for the final memories we leave to those who love us? The dignity that we seek in dying must be found in the dignity with which we have lived our lives. *Ars moriendi* is *ars vivendi*: The art of dying is the art of living. The honesty and grace of the years of life that are ending is the real measure of how we die. It is not in the last weeks or days that we compose the message that will be remembered, but in all the decades that preceded them. Who has lived in dignity, dies in dignity. William Cullen Bryant was only twenty-seven years old when he added a final section to his contemplation on death, "Thanatopsis," but he already understood, as poets often do:

> So live, that when thy summons comes to join
> The innumerable caravan, which moves

To that mysterious realm, where each shall take
His chamber in the silent halls of death,
Thou go not, like the quarry-slave at night,
Scourged to his dungeon, but, sustained and soothed
By an unfaltering trust, approach thy grave,
Like one who wraps the drapery of his couch
About him, and lies down to pleasant dreams.

Index

Vintage
Books